DATE DUE

SE 30 '97			
NO 4 '97			
DEC 08 '97			
MR 10 '98			
RENEW	JUN 8 '06		
31 '98	DE 21 '07		
MY 5 '98			
MY 26 '98			
MR 30 '98			
DE 18 '98			
FE 16 '99			
MR 9 '99			
AP 19 '99			
MY 10 '99			
MY 27 '99			
NO 10 '99			

Introduction to Fire Prevention

BRADY

Introduction to Fire Prevention

Fourth Edition

James C. Robertson

BRADY/PRENTICE HALL, *Upper Saddle River, New Jersey*   07458

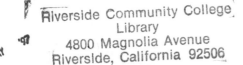

Library of Congress Cataloging-in-Publication Data

Robertson, James C.
   Introduction to fire prevention / James C. Robertson. — 4th ed.
      p.   cm.
   Includes index.
   ISBN 0-02-402241-1
   1. Fire prevention.   I. Title.
TH9145.R55   1995
363.37'7—dc20   94-25423
                                                            CIP

Editor: Susan Katz
Project Manager: Kimberly Kim
Cover Designer: Rose Marie Votta

   © 1995 by Prentice-Hall, Inc.
          A Simon & Schuster Company
          Upper Saddle River, New Jersey 07458

Printed in the United States of America

10 9 8 7 6 5 4 3 2

0-02-402241-1

Prentice Hall International (UK) Limited, *London*
Prentice Hall of Australia Pty. Limited, *Sydney*
Prentice Hall Canada Inc., *Toronto*
Prentice Hall Hispanoamericana, S.A., *Mexico*
Prentice Hall of India Private Limited, *New Delhi*
Prentice Hall of Japan, Inc., *Tokyo*
Simon & Schuster Asia Pte. Ltd., *Singapore*
Editora Prentice Hall do Brasil, Ltda., *Rio de Janeiro*

Dedicated to the memory of
Professor R. J. Douglas
Oklahoma State University

Contents

Chapter 6.   Enforcement of Fire Safety Codes   101

Chapter 7.   Programs for Public Fire Safety Education   119

Chapter 11.   Fire Prevention through Arson Suppression   217

Chapter 12.   Fire Prevention Research   240

Preface

Modern fire services emphasize fire prevention and fire safety aware-ness as an integral part of their responsibilities. The fourth edition of *Introduction to Fire Prevention* has been written with the goal of out-lining important elements to prevent fires, minimize losses, and engender appropriate reaction should a fire occur. Suppression of arson is included as a major fire prevention method.

Administrative and educational aspects of fire prevention and fire safety programs in both private and public sectors are addressed in *Introduction to Fire Prevention*. Historical aspects and the avenues fol-lowed in reaching the present state of the art are covered. The text does not include mechanical descriptions of automatic fire suppression and detection systems and devices, as other publications provide this information.

This edition has been completely updated to reflect changes in agency and organizational responsibilities. Chapters have been rearranged in a more logical sequence and subject matter no longer applicable has been deleted.

Acknowledgments

This revision represents the garnering of information from many sources. George Braun, Sheila P. Hayes, and Jackie D. Herndon, former associates in the Gainesville, Florida, Fire-Rescue Department; Anthony R. O'Neill, Vice President, Govern-ment Affairs, National Fire Protection Association; and John B. Woodall, Deputy Insurance Commissioner, State of Washington, were especially helpful with specific chapters. Many others also contributed updated information.

Rex Jordan, Fire Marshal, Westinghouse Hanford, Washington, Fire Department; and Thomas E. Kiurski, Adjunct Faculty, Oakland Community College, Southfield, Michigan, provided comprehensive reviews. Louanne Dryer Elliott, copyeditor, and Kimberly Kim, production supervisor, were also quite helpful. The assistance of my wife, Polly, who prepared the revisions for the publisher, is greatly appreciated.

J.C.R.

CHAPTER 1

History and Philosophy of Fire Prevention

Historical Background

A great deal can be learned by a study of the history of the development of fire prevention. It is hoped that the brief review given here will assist the reader in recognizing the reasons for certain procedures being followed in the field today.[1]

. As early as 300 B.C., the Romans established a "'fire department," composed primarily of slaves. The response of those individuals is reported to have been quite slow. Little else is known about their procedures; however, the program apparently was so unsuccessful that it was necessary to convert the department into a paid force in 6 A.D.

This apparently proved successful, and by 26 A.D. the full-time fire force in Rome had grown to approximately 7,000. These individuals were charged primarily with a responsibility for maintaining fire prevention safeguards. The population of Rome at the time was in the neighborhood of just under one million.

The fire brigades of Rome patrolled the streets in their efforts to bring about proper fire prevention procedures. They were granted the authority to administer corporal punishment to violators of fire codes and were provided with rods for use in administering such punishment. Records indicate that they were empowered to administer corporal punishment because "most fires are the fault of the inhabitants." It is difficult to imagine public acceptance today of corporal punishment in connection with fire prevention code enforcement. As an interesting sideline, in addition to their fire prevention duties, these patrolmen had the responsibility of keeping a watchful eye on the clothing of individuals who were using the public baths and were required to make inspections of the baths on a regularly scheduled basis.

In 872, according to history, a bell was used in Oxford, England, to signal the time for the extinguishment of all fires. From this appropriate time for "covering" of fires developed the Norman words *couvre feu* (cover fire), which later became the English word *curfew*. Records from 1066 indicate that during their occupation of England the

Normans strongly enforced the requirements for extinguishing fires at an early hour in the evening. Because construction at the time was such that dwelling fires could easily spread, this preventive measure was an effective safeguard.

As an added means of fire prevention, certain building code requirements were also imposed. Fitzstephen, writing in 1189 during the reign of Henry II, stated, "The only plagues of London were immoderate drinking by idle fellows and often fires."[2] This was an indication of a severe fire problem in that nation in those early days.

In an effort to control the fire situation, the lord mayor of London in 1189 issued an order to the effect that "no house should be built in the city but of stone and they must be covered with slate or tiled." This requirement was apparently vigorously enforced in structures built from that day on in the city. However, the full effect of that order was not yet felt by 1212, when the fire that caused the greatest loss of life to that date—it took 3,000 lives—occurred in London.

As a further example of the effort to reduce fire spread possibilities, Oxford in 1190 imposed a requirement for fire walls to be placed between every six houses. The purpose of this was to enable more rapid control of a fire and limit its spread.

Emphasis was on fire prevention in the Scottish Act of 1426. It ordered, for example, that "no hemp, lint, straw, hay or heather or broom be stored near a fire." Edinburgh merchants selling such wares were permitted to use lanterns, but not candles, and citizens in general were forbidden to carry naked flames from house to house.

Fire precautions figured largely in the Edinburgh Improvement Act of 1621. This act ordered noncombustible roofs and required tradespeople who kept "heather, broom, whins and other fuel" in the center of town to remove the material to remote areas.

High-rise buildings were also a problem in 17th-century Edinburgh. Some buildings reached 14 stories, which caused the Scottish Parliament to issue a regulation in 1698 restricting new buildings to a height of 5 stories. The regulation did not, however, affect existing buildings.[3]

History also records problems with arson during riots in the early days of England. In 1272 in Norwich, 34 rioters involved in arson and looting were captured and punished. Their punishment consisted of being dragged about town until dead. One woman arsonist was burned alive as punishment for her act.

In a later arson case, the punishment was equally severe. A 15-year-old male in Edinburgh who was judged responsible for setting fire to peat stacks was burned alive in 1585 as punishment for his act.

Specific punishments for fire prevention violations are also noted in historical documents. The city records of Southhampton, England, contain a late 1500s case in which a baker was fined two shillings for having combustibles too close to an oven. Manchester bakers were forbidden by a 1566 law to keep gorse (barley) "within two bays of the ovens."[4]

Charles II in 1664 gave authority for imprisoning those who contravened building regulations. These regulations related, then as now, to fire safety.[5] A 1763 act prohibited the piercing of fire walls.[6]

Among fire prevention recommendations issued to the public during those early days in England was one in 1643 that suggested that candles be placed in water-base holders. The thought was that an unattended candle would burn down and go out before causing trouble. An act of Parliament in 1556 required bellmen to patrol the streets and cry out, "Take care of your fire and candle."[7]

Fire alarm systems employed by communities in England were unique when compared with present practices. In Nottingham, a patrol of 50 women roamed the streets during the night in the interest of fire prevention. Their duty, in addition to detecting fires, was to awaken the sleeping community in the event of a fire.

· During 1666, a major fire struck London. This fire, referred to as the Great Fire of London, burned for four days and destroyed five-sixths of the city, yet only six deaths occurred. This was in sharp contrast with the 3,000 deaths that occurred during the fire of 1212. The effectiveness of the previously imposed fire prevention requirements undoubtedly had a bearing on the reduced number of deaths. Although thousands of structures were destroyed, the progress of the fire was retarded long enough to allow the occupants to vacate their premises.

As a further indication of efforts in the fire prevention field, an English citizen named David Hartley secured a patent for a fire prevention invention in 1722. The invention consisted of steel plates with dry sand between them intended to be used as a means of reducing fire spread from one floor to another, a method of construction that was employed in some of the early buildings in England. Mr. Hartley's invention was considered of sufficient value to have a statue erected in his honor.[8]

In 1794, theater fire protection was given a boost by the placement of a water tank on the roof of a theater in England. The tank provided a curtain of water in the event of a fire. In addition, an iron safety curtain was provided to separate the theater patrons from fire on the stage.[9]

The Birmingham, England, fire brigade issued a requirement in 1884 that inhabited tall buildings be provided with two staircases. This was looked on as a progressive fire protection measure.[10]

Early Fire Prevention Measures in North America _____

· Massachusetts Governor Winthrop, as a result of a serious fire in · Boston, issued an order in 1631 prohibiting the building of chimneys of wood and the covering of houses with thatch.[11]

Fire inspections in the New World probably began with the appointment by Governor Peter Stuyvesant in 1648 of four fire wardens to inspect wooden chimneys of thatched roof houses in New Amsterdam (New York City). These individuals were empowered to impose fines for improperly swept chimneys.[12]

Salem, Massachusetts, in 1644 imposed a fire safety ordinance requiring inhabitants to procure ladders for their houses. This was followed by a 1663 ordinance requiring that chimneys be swept each year.[13]

· Chimneys were a major fire problem in Colonial America. Philadelphia found it necessary in 1696 to prohibit the cleaning of chimneys by burning them out. Citizens were not allowed to smoke on the street by day or night, and the possession of more than six pounds of gunpowder within "forty paces of any building or dwelling" was prohibited.[14]

Norfolk, Virginia, prohibited wooden chimneys in 1731.[15] Easton, Maryland, in a 1791 ordinance, required chimneys to be built of brick or stone.[16] In 1796, New Orleans, then a Spanish province, passed an ordinance against the use of wood roofs.[17]

Rhode Island's first fire prevention law was enacted in 1704. It banned the setting of fire "in the woods in any part of this colony on any time of the year, save between the tenth of March and the tenth of May annually nor on the first or seventh day of any week." A subsequent measure enacted in 1731 prohibited unauthorized bonfires.[18]

Fire prevention enforcement measures were initiated in many communities during the early days of our country. As an example, in 1785 a city ordinance in Reading, Pennsylvania, imposed a fine of 15 shillings for each chimney fire that occurred in the city.[19] The fine was collected by the city and turned over to the fire company that had responded to the alarm. This ordinance was later repealed. Another Reading requirement was the alteration of chimneys in blacksmiths' shops to make them fire resistive, with a fine of $20 for violation.[20]

An 1807 ordinance in Reading prohibited the smoking of cigars on the street after sunset.[21] It also prohibited people from sitting on porches or in the doorway of any house with a lighted cigar or pipe without the consent of the owner. A $1 fine was imposed for violations of this ordinance. Under the Reading ordinance, the use of firecrackers was prohibited, with a fine of $1 or 12 hours in jail for violators. A duty was imposed on the citizens of Reading to confiscate and destroy fireworks found in the possession of a child.

The board of aldermen in Pensacola, Florida, passed an ordinance in 1821 requiring chimneys to be kept swept. A $10 fine was levied against the owner of any house whose roof caught fire.[22]

Jamestown, New York, imposed fire prevention regulations in 1827. Fire wardens were required to examine all chimneys, stoves, and other fire places used within Jamestown to direct "such reasonable repairs, cleansings, removals or alterations as shall be in his or their opinion best calculated to guard against injury by fire." Fines were imposed for failure to comply or for refusal of entry to the warden.

Occupants of shops or other places in Jamestown where rubbish might accumulate were required to remove accumulations as often as the warden saw fit. Fines were imposed for each day the violation continued.[23]

The first fire safety ordinance in Greensboro, North Carolina, enacted in 1833, required each household to have two ladders on its premises, "one which shall reach from the ground to the eaves of the house, the other to rest on top of the house, to reach from the comb to the eaves." Two inspectors were appointed to enforce this requirement and to ensure that all rubbish and nuisances were cleared from backyards. A five-dollar fine was imposed for each violation.[24]

In most newly formed towns, fire suppression forces were organized before the advent of fire prevention efforts. However, in Auraria, the original section of what is now Denver, Colorado, the legislative council in 1860 appointed six fire wardens "to inspect buildings and their chimneys and to prevent the accumulation of rubbish" as a result of burning of a large livery barn. The first fire fighting company was formed in 1866.[25]

More comprehensive fire prevention regulations were imposed in New York City in 1860 subsequent to a tenement building fire in which 20 people were killed. This ordinance required all residential buildings built for more than eight families to be equipped with fireproof stairs or fire escapes.[26]

Several major fires occurred in the early 1800s in Montpelier, Vermont. As a result of these fires, "the village appointed a committee of three to report a code of by-laws for the preservation of buildings from fire. The bailiffs were required to inspect every house in their ward to see that there was no fire hazard and that each place had, as required by by-laws, a fire bucket and ladder. Another by-law required that no fire should be left burning in a house unoccupied between the hours of 11:00 P.M. and 4:00 A.M., if adjacent to another."[27]

Fire escapes and exits attracted the attention of the Boise, Idaho, city council in 1887, when they imposed a requirement that doors on halls in theaters be made to swing outward. They were concerned about the possibility of a disaster at a performance in one of the city's places of assembly.[28]

Fire alarms and fire escapes, of course, had been invented, but they were not yet generally accepted. In fact, the Illinois legislature attempted to enact a fire escape law in 1897, one which would have replaced earlier, ineffective legislation. The 1897 act required fire escapes in all buildings more than four stories high and in all buildings higher than two stories if these structures were used as manufacturing places, hotels, dormitories, schools, or asylums. According to the *Centennial History of Illinois,* this act was bitterly fought by the Manufacturers' Association of Illinois. When passed, it proved impossible to enforce, and it was repealed in 1899. As late as 1912, 308 fire deaths were reported in Illinois, slightly less than half of these occurring in Cook County alone. Most victims were trapped in burning buildings. This suggests a continuing problem with fire exits and escapes, although the circumstances of the deaths were not individually reported.[29]

An 1896 fire that destroyed a saloon and hotel brought about the first fire prevention code in West Palm Beach, Florida. The ordinance established a fire district in which no building could be erected unless it was of brick, brick veneer, or stone construction.[30]

Captains of steam fire engine companies in Memphis, Tennessee, as early as 1900 performed inspections to locate and correct rubbish conditions in buildings, dangerous stove pipes, obstructed fire escapes, and defective chimneys and flues. Great amounts of cotton were stored on vacant lots and on streets, which further contributed to the fire problem.[31]

Formal fire prevention measures in Tulsa, Oklahoma, apparently began with a 1906 requirement that owners of all buildings with three or more stories install fire escapes. Failure to comply by a set date resulted in a fine of $15 per day. Store owners were prohibited from using rubber tubing for gas connections. Failure to comply resulted in steel piping being installed at the expense of the owner.[32]

Fire chiefs in the United States have long had an interest in fire prevention. The first general topic discussed at the First Annual Conference of the National Association of Fire Engineers (predecessor to the International Association of Fire Chiefs) was fire prevention. The following specific fire prevention items were discussed at the conference, which was held in Baltimore in 1873:

I. The limitation or disuse of combustible materiel in the structure of buildings; the reduction of excessive height in buildings and the restriction of the dangers of elevator passages, hatchways, and mansards.

II. The isolation of each apartment in a building from other apartments, and of every building from those adjoining by high party-walls.

III. The safe construction of heating apparatus.

IV. The presence and care of trustworthy watchmen in warehouses, factories, and theatres, especially during the night.

   V. The regulation of the storage of inflammable materiel, and the use of the same for heating or illumination; also exclusion of rubbish liable to spontaneous ignition.

   VI. The most available measures for the repression of incendiarism.

   VII. A system of minute and impartial inspection after the occurrence of every fire, and rigid inquiry into the causes, with reference to their future avoidance.

   VIII. Fire escapes actually serviceable for invalids, women, and children.[33]

The use of fire suppression personnel for prefire planning inspections was discussed by the Salt Lake City fire chief at the 1901 conference. At their conference in 1902, fire chiefs discussed developments in fire-retardant paint and slow-burning wood.[34]

In Milwaukee, fire prevention requirements were first imposed for places of public assembly in 1888. Violations of these regulations carried fines of $5 to $100. Apparently, these were the only regulations of a fire prevention nature in effect in the community at that time. By 1913, Milwaukee had a force of 30 men devoted entirely to fire prevention duties. There were 90,000 buildings in the city at that time. The personnel assigned to fire prevention duties were paid entirely through the returns from an insurance premium tax. By 1919, more than 250,000 inspections were being conducted each year by the fire prevention bureau.[35]

Fire prevention bureaus were started after 1900 in a number of larger cities. Long Beach, California, established such a bureau in 1917, while Phoenix, Arizona, started theirs in 1935. At that time Phoenix had a population of 46,500.[36,37]

The development of water distribution systems has played a major role in community fire defense. In Houston, Texas, the first fire engine arrived in 1839; however, a public waterworks did not come about for many years. By the mid-1870s, most businesses had cisterns for fire protection. In late 1878 the city entered a contract for the development of a water distribution system which was in service by the following summer. This pattern of water system development is typical of North American cities.[38]

Canada's first fire prevention regulation, enacted in Quebec City in 1692, required householders to clean out chimneys at prescribed intervals. This city was the site of the nation's first large loss of life fire. The Theatre Royale fire in 1846 took 47 lives.[39]

The development of fire safety regulations in Toronto was described in the history of that city's fire department as follows:

> When man realized he could gain more living space on the same sized lot by adding floors on top of each other, city council paid more attention to the types of construction, the principal consideration being to guard

against fire. Regulations were comprehensive and strict. A code in 1890 indicated that buildings were not to be erected on streets less than 35 feet wide or without a vacant space of at least 300 square feet. Furnaces, stoves and chimneys had to meet certain standards. Ladders had to remain in place in the same manner as indicated by the law in 1820. Inflammable liquids and gunpowder were restricted. Churches, hospitals, colleges, schools, halls and theatres must have stairs and doors as required by the director of buildings. No seats were to be put in the aisles. Theatres had to have fire hoses plugged in. The city was divided into four fire limits, the first being approximately bounded by Jarvis, Queen and Simcoe Street and the Esplanade. That limit was to have maximum protection including fireproof walls and roofs.[40]

In Calgary, Alberta, the fire chief first recommended the creation of a fire prevention bureau in his report to the mayor and aldermen in 1912. This request was included with ever-increasing emphasis in subsequent years until the bureau was finally established in 1951. In the meantime, the fire chief was making some inspections within the city; however, no full-time force was available for such efforts.[41]

Tragedy a Spur to Regulations _____

It has been said that "in the realm of fire 'the law' is a thing mothered by necessity and sired by great tragedy."[42] How true this is will be evident from a brief review of some of the major fires that have occurred through the years and their relation to the development of fire safety regulations and procedures in the United States and Canada.

Public Assembly

On December 5, 1876, a major fire occurred in the Brooklyn Theater in New York. In this fire a stage backdrop was ignited, and 295 people were killed under conditions similar in many ways to those in the Iroquois fire 27 years later.[43]

The Iroquois Theater fire in Chicago, outstanding among public assembly fires, occurred in 1903 during a Saturday matinee of a new play, *Mr. Bluebeard.* The Iroquois was Chicago's newest theater and was also considered its safest; it was advertised as being "Absolutely Fireproof." There were 2,000 people present for this performance. Arc lamps were used in the theater. A light set a curtain on fire, and flames and smoke rapidly made the structure untenable. Despite heroic efforts, panic ensued, and human logjams developed at each of the doors. No fire extinguishers were provided. The curtains were combustible, and exits were improperly marked and swung inward. No venting was provided for the stage area; so there was no means of immediately removing hot gases and smoke. This tragic fire took 603 lives and gave a great deal of impetus to the fire prevention movement, especially in the field of public assembly occupancies.

In Natchez, Mississippi, a fire in a small dance hall, the Rhythm Club, took 207 lives and caused injuries to 200 more in 1940. Combustible decorations and one exit with the door opening inward were the factors responsible. Over 700 patrons were in the one-story building, which measured only 120 feet by 38 feet.

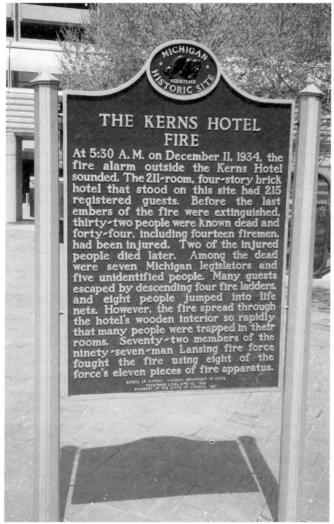

Figure 1-1   Many high-fatility fires are memorialized only at the local level.

During the early days of World War II, a major fire struck the Cocoanut Grove nightclub in Boston, Massachusetts. On the night of the fire, November 28, 1942, the club had approximately 1,000 occupants, many of whom were people preparing to go overseas on military duty. A lighted match used by an employee in changing a light bulb has been considered the possible cause for this tragic fire, which took 492 lives. Almost half of the occupants were killed, and many were seriously injured. Flammable decorations spread the fire rapidly. Men and women were reported to have clawed inhumanly in an effort to get out of the building. The two revolving doors at the main entrance had bodies stacked four and five deep after the fire was brought under control. Authorities estimated that possibly 300 of those killed could have been saved had the doors swung outward. It should be noted that the capacity of the structure had also been exceeded.

The Cocoanut Grove fire prompted major efforts in the field of fire prevention and control for nightclubs and other related places of assembly. Immediate steps were taken to provide for emergency lighting and occupant capacity placards in places of assembly. Exit lights were also required as a result of the concern generated by this fire.

Fire protection under the big top was emphasized by the fire that struck the Ringling Brothers, Barnum and Bailey Circus while playing in Hartford, Connecticut, on July 6, 1944. Seven thousand people attended the daytime performance. The circus tent, which measured 425 feet by 180 feet, was apparently not properly flame-retardant, and the fire caused 163 deaths and 261 injuries. After this fire, many states and municipalities gave more attention to circus fire safety requirements. It is ironic that the fire occurred in Hartford, a city that had had an outstanding fire prevention program for many years.

On May 28, 1977, a tragic fire struck the Beverly Hills Supper Club in South Gate, Kentucky. At the time of this fire, which took 164 lives, the club was occupied by 3,000 to 3,400 people. The building, which had an area of 54,000 square feet, was of unprotected, noncombustible construction. Fire separations, automatic sprinklers, and other safeguards were lacking. Exits were insufficient for the capacity crowd. Interior furnishings were made of combustible materials.

The Beverly Hills fire spurred new demands for improved fire safety measures, including inspection improvements. Many of the patrons in the club at the time of the fire were from other jurisdictions that strongly enforced codes for public assembly occupancies. National political leaders raised the question of the propriety of citizens of one jurisdiction being exposed to fire danger when visiting an area where code enforcement is not as stringent. The impact of this fire will be felt for many years to come.

Industrial Facilities

Another fire that made an impact on structural fire safety regulations was the fire that occurred in the Triangle Shirtwaist Factory in New York City on March 25, 1911. More than 600 women, most of them young, were working on the eighth, ninth, and tenth floors of this loft building. To prevent unauthorized removal of products, the factory management made a practice of checking purses and bags of all employees as they left the premises. During this procedure the exit doors to the stairs were locked.

The fire started from an unknown cause on the eighth floor at approximately 4:45 P.M. Interior standpipe hose was rotten and completely ineffective. The fire spread out the windows and onto the floors above. Sixty people jumped to the ground from the eighth, ninth, and tenth floors; a total of 146 people were killed in this disaster. In addition, 70 people were injured, many for life. The owner of the factory and his family, who happened to be in the building, escaped by going up through the floor above. Charges were brought against the owner as a result of the investigation; however, he was acquitted because he apparently proved that he did not know about the practice of locking the stairway exit doors during the search for merchandise being removed from the premises.

This fire focused attention on the need for fire safety measures and the safeguarding of occupants in similar buildings. The nation was shaken by this incident, which led to regulations to preclude the possibility of a recurrence. The fire led to enactment of New York State's Labor Law and establishment of the Fire Department's Bureau of Fire Prevention.[44]

Although no loss of life ensued, the General Motors Transmission Plant fire in Livonia, Michigan, on August 12, 1953, had a major impact on industrial fire safety. This fire, in a building with an undivided floor area of 34.5 acres with only partial sprinkler protection, resulted in the greatest damage from an industrial fire as of that date.

On April 16, 1947, a major disaster struck the waterfront of Texas City, Texas. The S.S. *Grandcamp* was taking on a shipment of ammonium nitrate fertilizer at the pier of the Monsanto Chemical Co. A small fire was discovered aboard, but before it could be brought under control, the ship blew up, instantly killing almost the entire fire department of Texas City, which had responded to the first alarm of fire, as well as all but seven of the ship's crew. Fire and other explosions that followed during the ensuing hours in the waterfront industrial area resulted in the deaths of 468 people, over 2,000 injuries, and property loss of over $67 million. This disaster emphasized the need for regulations in the control of fertilizer-grade ammonium nitrate.

The ports of the nation and the U.S. Coast Guard began vigorous procedures to eliminate the possibility of having vessels containing such materials within their limits.

Halifax, Nova Scotia, was devastated by the collision of two ships, one loaded with ammunition, in the harbor on December 6, 1917. The ensuing explosion and fires resulted in 1,960 deaths, 9,000 injured, 6,000 homeless, and the destruction of 1,600 buildings and damage to thousands more.

Fire struck the Imperial Foods Processing Plant in Hamlet, North Carolina, on September 3, 1991. The fire's intensity, coupled with several inoperable exits, resulted in 25 fatalities and 54 injuries in this unsprinklered one-story, windowless structure.

Hotels

Nationwide interest in hotel fire safety was awakened by the fire that swept the Winecoff Hotel in Atlanta, Georgia, on December 7, 1946. This building was widely advertised as a "fireproof" structure and was, in fact, of fire-resistive construction. An open stairway, however, permitted the rapid spread of smoke and heat up the stairs from floor to floor. There were 304 guests in this 15-story building the night of the fire; 119 people were killed, 168 injured. A number of occupants jumped from windows.

Reports on this fire indicate that some people who stayed in their rooms were not injured by the fire. They were protected by room doors and managed to obtain air for breathing by opening the windows. Great strides in hotel fire safety came about as a result of this fire. Fire safety improvements were made within the entire state of Georgia, which adopted a fire code soon after the fire.[45]

During the same year, a fire at the La Salle Hotel in Chicago took 62 lives, and 19 lives were lost in a fire in the Canfield Hotel, Dubuque, Iowa. Many similarities existed in the fires in the three hotels.

A 1986 mid-afternoon New Year's Eve fire struck the DuPont Plaza Hotel and Casino in San Juan, Puerto Rico. The fire resulted in 96 fatalities and over 140 injuries. This 20-story, nonsprinklered hotel contained a first-floor ballroom, a second-floor casino, and various mercantile shops, restaurants, and conference rooms. Fire consumed the contents of first-floor areas and was primarily confined to that level. Investigation revealed that the fire had been deliberately set in a large stack of recently delivered furniture that had been temporarily stored in the ballroom.

The government of Puerto Rico convened a commission to study the need for improvements in fire safety code. The commission report brought about improved fire safety standards.

A fire in the MGM Grand Hotel in Las Vegas, Nevada, on November 21, 1980, took 85 lives and injured 600. This fire began in the casino area; however, most of the victims were in the adjacent high-rise section.

Nursing Homes, Hospitals, and Housing for the Elderly

A number of serious fires have occurred in nursing homes throughout the country. One that focused attention on the field of fire protection and prevention was the Katie Jane Nursing Home fire in Warrenton, Missouri, in 1957. The fire occurred on a Sunday afternoon, when many visitors were in the home, and took 72 lives. Construction deficiencies, the lack of an automatic sprinkler system, and a number of other factors were paramount in this disaster. However, this fire and one that took 63 lives in a Fitchville, Ohio, home for the aged in 1963 did result in the improvement of nursing-home fire safety regulations and procedures in a number of jurisdictions. Until corrective action was initiated at the federal level early in the 1970s, a study of fires indicated a continuing pattern of disastrous fires in nursing homes and related institutions.

Fire safety problems in hospitals were brought to national attention by the St. Anthony's Hospital fire in Effingham, Illinois, on the night of April 4, 1949, in which 74 lives were lost. Combustible interior finish, the lack of automatic sprinkler protection, and general construction were factors in this fire.

A fire that occurred in 1961 in the Hartford Hospital in Hartford, Connecticut, was responsible for 16 deaths. Combustible ceiling tile contributed to the fire's spread, as did the fact that trash and linen chutes in the building opened directly onto the corridors. The structure was a modern fire-resistive building, which proved that such structures are not immune to fires.

In just 13 days in December, 1989, fires killed 23 residents in three different housing-for-the-elderly facilities. None of these were licensed or operated as nursing homes.

The first occurred in a six-story, fire-resistive, unsprinklered apartment house for the elderly in Roanoke County, Virginia. Four residents died on the third floor, although the fire was confined to the room of origin. The investigation revealed that some of the occupants were difficult to awaken and reluctant to leave their apartments.

Watertown, New York, was the scene of the second fire. Three fatalities occurred, all on the upper floors of this nonsprinklered, fire-resistive apartment building constructed specifically for housing the elderly. Smoke movement from the first-floor point of origin was responsible for the fatalities in this December 15 late-night fire.

On December 24, a late-afternoon fire started on the first floor of an 11-story fire-resistive apartment house in Johnson City, Tennessee. This nonsprinklered structure had served as a hotel for many years before conversion to housing for the elderly. Sixteen fatalities and 40 injuries occurred. As in the other fires, combustible interior finish was a factor in this incident. Most of the fatalities were on upper floors.

Generally, fire safety requirements in apartments for the elderly are no more stringent than those for any other apartment house. Special licensure is normally not required.[46]

Schools

A six year-old, three-story school in Colinwood, Ohio, was the scene of tragedy in 1908. The Lakeview School was occupied by over 300 students at the time of a fire. Of those students, 175 were killed. Open stairways and construction materials contributed to the fire spread.

An early school fire brought about a realization that fire safety should be considered even in the smallest of schools. This fire occurred in a two-room elementary school at Babbs Switch, Oklahoma, on December 24, 1924. The school was being used for a Christmas event attended by over 200 community residents. A Christmas tree had been placed in the opening between the two rooms. The windows in the school had previously been barred, and there was but one exit. As a result of these unsafe conditions, 36 deaths ensued in this fire, which was started by a toppled candle on the tree.

In March 1937 a major explosion struck the New London, Texas, school. This disaster, which killed 297 people, was caused by the improper use of gas. Again, school fire safety was immediately considered to be of paramount concern in communities throughout the country.

School fire prevention and fire protection efforts were greatly enhanced by the tragic fire at Our Lady of Angels School in Chicago on December 1, 1958. This fire, in which 95 lives were lost, was reported to be of incendiary origin. The old multistory structure was of ordinary construction. A delay in turning in an alarm within the building and in contacting the fire department compounded the tragedy. Tremendous efforts to upgrade school fire prevention regulations throughout the country came as a result of this fire. Many schools were provided with automatic sprinkler protection. In addition, a number of states started vigorous fire inspection programs and took other measures to reduce the possibility of fires taking lives because of lack of safeguards.

Prisons

The Ohio State Penitentiary fire of April 21, 1930, in Columbus, aroused interest in problems relating to fire prevention in prisons. The penitentiary had 4,300 inmates at that time, and 320 were killed in the fire. Most of those killed were trapped in their cells. This fire brought about immediate demands for improvements in fire safety in penal institutions.

Three major prison fires killed a total of 68 people during 1977. Included were 5 fatalities in the federal prison in Danbury, Connecticut, 42 in the county jail in Maury County, Tennessee, and 21 in a St. John, New Brunswick, penal institution. These fires led to new demands for tighter regulation of fire safety within places of incarceration. Interior finish had become a major contributory factor to fires in such places.

Conflagrations

Until the 20th century, communities had been built to burn while the country expanded rapidly across the continent. Building conditions in the United States had long been marked by excessive use of combustible materials put together without much regard for safety to life or property from fire. Large individual buildings housing vast amounts of combustible stocks under one roof; lack of fire walls and vertical cutoffs; wood-shingle roofs; and other unsafe factors that contribute to rapid spread of fires were characteristic of the American scene. In many parts of the United States, seasonal droughts and high winds aggravated fire conditions and resulted in area conflagrations that contributed substantially to the high national fire losses.

October 9, 1871, was the day of the two greatest fires in the history of the Midwest. Both started at about the same hour and were compounded by tinder-dry wood resulting from a prolonged drought, plus human carelessness. The better known of the two struck Chicago, taking 300 lives, while structures in an area four miles long and two-thirds of a mile wide were consumed during a two-day period.

The lesser known fire ravaged 2,400 square miles and took 1,152 lives in a forested area of northeastern Wisconsin. This fire is recorded as the Great Pestigo fire because the majority of fatalities occurred in that Wisconsin town. The "Great Fire of 1901," which burned 146 city blocks in Jacksonville, Florida, on May 3, 1901, was typical of the conflagrations that struck U.S. and Canadian cities during that era. Only seven lives were lost in this fire, which destroyed 2,368 buildings. The fire began in a fiber factory and spread rapidly because of wood shingle roofs.[47]

Major conflagrations occurring in U.S. cities have also brought reforms in building and fire prevention codes. In 1904, Baltimore was swept by a conflagration that destroyed 80 blocks in the downtown business center. This fire resulted in the improvement of construction standards and the development of new procedures in fire prevention. Other North American cities and towns have had somewhat similar occurrences. For example, an 1861 fire destroyed most of the town of Lindsay, Ontario. Four hotels, two mills, the post office, customs office, and 83 other buildings were destroyed in the town of 2,000. A history of Lindsay states, "The fire spurred the construction of many fine brick buildings to replace the wooden structures which had been consumed."[48]

On October 12, 1918, forest fires swept across northeastern Minnesota, devastating 38 communities and 1,500 square miles. More than 450 people were killed outright, 52,000 more were injured or displaced, and property loss was estimated at $73 million.[49]

In 1923, Berkeley, California, was ravaged by a fire that destroyed 640 structures. In 1961 a conflagration in Los Angeles spread through wooded lands and resulted in the destruction of over 500 dwellings. Some large communities permit the use of wood-shingle roofing. The Los Angeles fire emphasized the need to return to more restrictive code provisions regarding wood-shingle roofing.

Oakland, California, was the site of a devastating fire that took 26 lives and resulted in the loss of 3,469 housing units on October 20, 1991. Property losses were estimated at $1.5 billion. Dry conditions and steep terrain contributed to this loss.

Earthquakes and Fire

Although not directly controllable under normal fire safety procedures, the 1906 earthquake and fire in San Francisco must be mentioned in any recount of fires of major significance. The severity of the fires was much greater because of the earthquake. Water mains were broken, which greatly reduced the amount of water available for fire fighting purposes. In addition, the earthquake made response by fire fighting equipment most difficult. A number of fires were uncontrolled. In this disaster, 422 lives were lost, 28,000 buildings were destroyed, and property loss was estimated at $350 million. This devastating event prompted reconsideration of fire prevention procedures in all areas of the country in which earthquakes might be anticipated.

Laboratories Improve Fire Safety _____

Some of the most important fire prevention efforts start in the laboratory. An example is the work of Dr. Robert C. Kedzie of Michigan. In the early 1870s, residents depended on kerosene oil lamps for illumination. The oil often contained volatile impurities, which caused explosions followed by burnings and death when lamps were ignited.

Kerosene imported into Michigan bore a label "Warranted 150° F Fire Test." This was based on an open-cup test, which did not confine the kerosene vapor to the cup, thereby tending to result in false readings. Dr. Kedzie devised a closed cup test, which confined the vapors and provided a much more accurate test, because it simulated conditions in a kerosene lamp. In many cases the actual flash point was found to be below 100° F with this test. Mandatory use of the closed cup test has resulted in the saving of countless lives.[50]

Moreover, the foundation of the modern plastics industry—highly flammable products generally yielding toxic fumes—had been laid in the 19th century. The first such substance was Parkesine, made in England in 1866. It was composed of nitrocellulose, camphor, and alcohol. A similar material, celluloid, was manufactured in the United States in 1869. It quickly came into use in everyday items such as shirt collars. A form of this product was used for early motion picture films. Great clouds of yellow smoke billowed out over Ohio's Cleveland Clinic in 1929, when burning nitrate celluloid film plates spread noxious fumes throughout the clinic, leaving 125 dead. Most were killed by the fumes rather than by the fire that followed. Safety film was developed soon thereafter.[51]

Another example of improvements attempted through the laboratory is the effort to obtain a fire-safe cigarette. The first American patent for a "self-extinguishing" cigarette was issued in 1854. Research to prove the practicability of such a cigarette began in the then National Bureau of Standards in 1932 and has continued through the years as the result of legislative action. Canadian interest in this problem has paralleled that in the United States.

Variables in the Philosophy of Fire Prevention _____

The philosophy of fire prevention includes many variables. The term *fire prevention* varies in interpretation within the fire protection field. There had been a tendency to include some closely related activities under fire prevention responsibilities. Some of these activities are not truly preventive in nature.

As an example of added meanings, a number of activities that are actually *fire reactions* are considered to be "fire prevention practices" by both lay and professional people. The practice of home fire drills, for example, is usually associated with fire prevention programs even though it is actually a fire reaction type program. Probably one reason for this is that the same individuals generally promote both programs.

Certain fire prevention concepts are not strictly related to the prevention of fire, but are more closely related to the *prevention of the spread of fire*. For example, wearing noncombustible clothing will not prevent the ignition of a match, but wearing such clothing does retard the possible spread of a fire that might be started on the clothing by the match. The personal steps taken to reduce fire spread possibilities on clothing are one aspect of fire prevention. But measures taken to preclude the possibility of starting a fire are more truly the essence of fire prevention.

Another fire prevention measure is making sure that fires that might occur in a structure will not entrap individuals within the building. People who happen to be in the structure at the time of the fire should have every opportunity to leave the building safely. An important factor here is individual conditioning to fire reaction, in which health, physical abilities, past exposure to fire, and many other variables play a part. The term *fire safety* is seen by many as encompassing fire prevention, fire reaction, and prevention of fire spread.

Summary _____

The dangers of fire to life and property are ever present, and from earliest times efforts have been made to prevent fire.[52] In ancient Rome, firemen patrolled the streets to enforce fire prevention procedures and administered corporal punishment to offenders of fire codes. Records of fire prevention efforts in England date back to 872 and reveal some unusual practices: a bell was rung as a signal to extinguish all household fires at an early evening hour; women patrolled the streets at night to detect fires and to act as a fire alarm system; arsonists were punished by being burned alive.

Fire prevention efforts in our country began as early as 1631. In 1785 and 1807, ordinances enacted in Reading, Pennsylvania, imposed fines on homeowners for chimney fires, and prohibited the smoking of cigars on the street after sunset and the firing of firecrackers. Fire prevention was the first general topic discussed at the First Annual Conference of the National Association of Fire Engineers (predecessor to the International Association of Fire Chiefs), which was held in Baltimore in 1873.

From those modest beginnings, fire prevention has come to be recognized as a science. But in practical application, major improvements have come about mostly as a result of tragic fires at the cost of many lives and extensive property damage. The list is long: the Iroquois Theater fire in Chicago, 1903, 603 killed; the Triangle Shirtwaist Factory, New York, 1911, 146 deaths; the Cocoanut Grove nightclub, Boston, 1942, 492 dead, scores injured; nursing homes, hospitals, schools, conflagrations like the one in Baltimore in 1904, which destroyed 80 blocks in the downtown business center, and the one in 1991 in Oakland California, in which over 3400 dwellings were destroyed.

Notes

1. Historical references to early Roman and English periods drawn from *A History of the British Fire Service* by G. V. Blackstone, CBE, G.M. (London: Routledge & Kegan Paul, 1957).
2. Ibid., p. 10.
3. Alexander Reid, *"Aye Ready!"* (Edinburgh: Geo. Stewart & Co., 1974), p. 5.
4. Blackstone, p. 17.
5. Ibid., p. 33.
6. Ibid., p. 87.
7. *Project 9: The Story of Fire Fighting* (London: The Home Office and the Central Office of Information, 1976), p. 1.
8. Blackstone, p. 84.
9. Ibid., p. 87.
10. Harry Klopper, *The Fight Against Fire* (Birmingham, England: Birmingham Fire and Ambulance Service, 1955), p. 36.
11. Paul R. Lyons, *Fire in America!* (Boston: National Fire Protection Association, 1976), p. 2.
12. Charles L. Radzinsky, *100 Years of Service 1872–1972* (Rensselaer, N.Y.: Hamilton Printing Company, 1972), p. 9.
13. Edward R. Tufts, *A History of the Salem Fire Department* (Salem, Mass.: Holyoke Mutual Insurance Company in Salem, 1975), p. 3.
14. *History of the Fire Department, Norfolk, Virginia* (Norfolk, Va.: Norfolk Firefighters' Association, 1975), p. 32.
15. Ibid., p. 8.
16. James C. Mullikin, *A History of the Easton Volunteer Fire Department* (Easton, Md.: Easton Volunteer Fire Department, 1962), p. 9.
17. The National Commission on Fire Prevention and Control, *America Burning* (Washington: The National Commission on Fire Prevention and Control, 1973), p. 79.
18. Patrick T. Conley and Paul R. Campbell, *Firefighters and Fires in Providence* (Providence, R.I.: Rhode Island Publications Society, 1985), p. 3.
19. *Reading's Volunteer Fire Department,* comp. Federal Writers Project of Works Progress Administration in the Commonwealth of Pennsylvania, Berks County Unit (Philadelphia: William Penn Association, 1938), p. 6.
20. Ibid., p. 7.
21. Ibid.
22. Arnold Rosenbleeth, *Firefighting in Pensacola* (Pensacola, Fla.: Pensacola Historical Society Quarterly, Winter 1980), p. 2.
23. Leo E. Duliba, *A Transition in Red* (Merrick, N.Y.: Richwood Publishing Co., 1976), pp. 12–13.
24. *Greensboro Fire Department, 1808–1984* (Dallas, Tex.: Taylor Publishing, 1984), p. 6.
25. "Past and Present History Denver Fire Dept.," unpublished paper, Denver, Colo., p. 2.
26. Lyons, pp. 29–30.
27. Perry H. Merrill, *Montpelier, the Capital City's History* (Montpelier, Vt.: published by author, 1976), p. 62.

28. Arthur A. Hart, *Fighting Fire on the Frontier* (Boise, Idaho: Boise Fire Department Association, 1976), p. 33.
29. Betty Richardson and Dennis Henson, *Serving Together: 150 Years of Firefighting in Madison County, Illinois* (Collinsville, Ill..: Madison County Firemen's Association, 1984), p. 45.
30. *The Fire History of the City of West Palm Beach* (West Palm Beach Fire Department, Fla.: 1980), p. 6.
31. Frank and Gennie Myers, *Memphis Fire Department* (Marceline, Mo.: III Walsworth, 1975), p. 30.
32. *Tulsa Fire Department 1905–1973* (Tulsa, Okla.: Intercollegiate Press, 1973), p. 17.
33. Donald M. O'Brien, *The Centennial History of the International Association of Fire Chiefs* (n.p., 1973), p. 6.
34. *Ibid.,* p. 43.
35. R. L. Nailen and James S. Haight, *Beertown Blazes: A Century of Milwaukee Fire Fighting* (Milwaukee, Wisc.: NAPCO Graphic Arts, 1971), pp. 21–22.
36. *Long Beach Fire Department* (Long Beach, Calif.: Long Beach Fire Department, 1976), p. 29.
37. *Phoenix Fire Fighters* (Phoenix, Ariz.: Phoenix Fire Department, 1983), p. 23.
38. Houston Fire Museum, Inc., *Houston Fire Department 1838–1988* (Dallas, Tex.: Taylor Publishing, 1988), p. 13.
39. Donal M. Baird, *The Story of Firefighting in Canada* (Erin, Ontario: The Boston Mills Press, 1986), pp. 14, 53.
40. *History of the Toronto Fire Department* (Toronto, Ontario: Toronto Fire Department, 1984), p. 42.
41. Grant MacEwan, *100 Years of Smoke, Sweat, and Tears* (Calgary, Alb.: Calgary Fire Department, 1984), p. 57.
42. Bob Considine, *Man Against Fire* (Garden City, N.Y.: Doubleday and Co., 1955), p. 134.
43. *Great Fires of America* (Waukesha, Wisc.: Country Beautiful Corporation, 1973), p. 134.
44. *WNYF* (New York: New York Fire Department, 3rd issue, 1980), p. 20.
45. Sam Heys and Allen B. Goodwin, *The Winecoff Fire* (Marietta, Ga.: Longstreet Press, 1993).
46. *Alert* (Bulletin No. 90–1, Quincy, Mass.: National Fire Protection Association, 1990).
47. James Robertson Ward, *Old Hickory's Town* (Jacksonville, Fla.: Old Hickory's Town, Inc., 1985), p. 175.
48. Alan R. Capon, *Historic Lindsay* (Belleville, Ont.: Mika Publishing, 1974), p. 18.
49. Francis M. Carroll and Franklin R. Raiter, *The Fires of Autumn* (St. Paul, Minn.: Minnesota Historical Society Press, 1990).
50. George P. Merk, "George C. Kedzie, Michigan's Nineteenth Century Consumer Advocate," *Michigan History* (Lansing, Mich.: Bureau of History, Secretary of State, Jan.–Feb. 1989), pp. 20, 21.
51. Richardson and Henson, p. 56.
52. Blackstone, p. 17.

CHAPTER 2

Development and Enactment of Fire Safety Codes

Various types of codes have a bearing on fire prevention. At the municipal level, these include codes relating to fire, buildings, zoning, planning, electrical installations, mechanical equipment, plumbing, and other specialized subjects.

The fire safety or prevention code in a municipality is the primary subject of interest to this study, although the building code probably has more impact on overall fire prevention in the community. Fire prevention codes, which have historically been concerned with fire safety regulations that relate to fire protection equipment, maintenance of buildings and premises, and of hazardous materials, processes, and machinery used in buildings, are generally administered and enforced by the fire department.

Building codes, which are usually administered and enforced by the building department, address fire safety requirements with respect to the construction of buildings. Because there may be some areas of overlap, a cooperative relationship between the fire department and the building department is essential for the good of the community. It should be noted that neither fire prevention nor building codes address furnishings of dwellings, which are major contributors to fire fatalities.

Generally, fire safety and building codes, when adopted for the first time in a state or community, do not apply to existing structures or installations except when the enforcement official determines that continuation of the hazard would jeopardize safety of life and property. The inclusion of such a clause does not exempt existing facilities from code application; it merely places on onus on the enforcement official to be in a position to justify retroactive application from a life and property safety standpoint.

For many years the municipal fire prevention code was thought of as a "housekeeping" code that related to the actual use of a facility from a fire prevention standpoint. Limited consideration was given in some codes to structural fire protection measures such as egress facilities or automatic extinguishing systems. This situation is changing, however.

Fire prevention requirements in some form or another have been a part of our society for many generations. The inclusion of a document known as a fire prevention or fire safety code in the municipal government framework can be traced in more recent years to the influence of the insurance industry on public fire protection. The National Board of Fire Underwriters, in the aftermath of the great Baltimore fire of 1904, developed the Standard Grading Schedule[1] for cities and towns with reference to fire defenses. This had a profound effect on the development and enactment of municipal fire prevention codes.

The National Board of Fire Underwriters (now known as the American Insurance Association) also formulated a model municipal Fire Prevention Code,[2] copies of which were made available to municipalities for the asking. Countless jurisdictions throughout the Untied States adopted the suggested code without change as the legally constituted fire prevention code. Although some cities probably would have developed and implemented fire prevention codes on their own, it is doubtful that there would have been anywhere near the use of fire prevention codes without the availability of the model code.

Model Fire Prevention Codes _____

The National Fire Protection Association Fire Prevention Code covers subjects typical of those covered in a modern state, county, or municipal fire prevention code. This Code, which is adopted in a number of jurisdictions, has chapters covering administration and enforcement; general fire safety requirements; means of egress; fire protection systems and equipment; automatic sprinkler systems; fire detection and alarm systems; occupancy fire safety requirements, ranging from assembly occupancies to airports and heliports; and special processes and material handling ranging from hazardous materials and chemicals to safeguards during building construction and demolition operations. The Code also contains a definitions chapter and an appendix. As with the other model codes, updated editions are developed every few years.[3]

The model Fire Prevention Code of the American Insurance Association (formerly the National Board of Fire Underwriters) mentioned in the preceding discussion was for many years widely used by the nation's fire service. The Association no longer maintains this code.

The Uniform Fire Code was originally developed by Southern California fire inspectors and first adopted by the California Fire Chiefs Association in 1961. Subsequently, the Western Fire Chiefs Association, in cooperation with the International Conference of

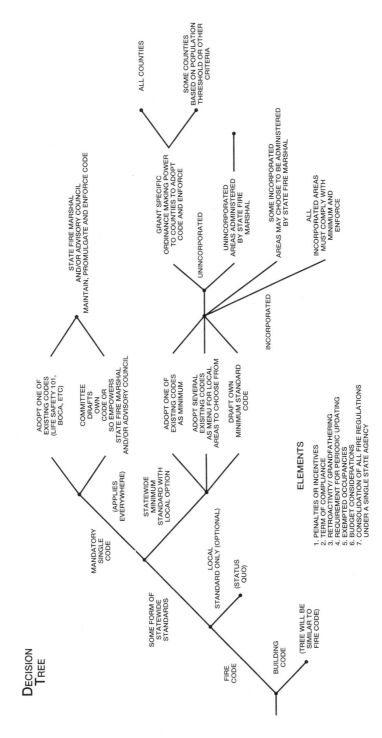

DECISION TREE

ALL COUNTIES

SOME COUNTIES BASED ON POPULATION THRESHOLD OR OTHER CRITERIA

STATE FIRE MARSHAL AND/OR ADVISORY COUNCIL MAINTAIN, PROMULGATE AND ENFORCE CODE

GRANT SPECIFIC ORDINANCE MAKING POWER TO COUNTIES TO ADOPT CODE AND ENFORCE

UNINCORPORATED

UNINCORPORATED AREAS ADMINISTERED BY STATE FIRE MARSHAL

SOME INCORPORATED AREAS MAY CHOOSE TO BE ADMINISTERED BY STATE FIRE MARSHAL

INCORPORATED

ALL INCORPORATED AREAS MUST COMPLY WITH MINIMUM AND ENFORCE

ADOPT ONE OF EXISTING CODES (LIFE SAFETY 101, BOCA, ETC)

COMMITTEE DRAFTS OWN CODE OR SO EMPOWERS STATE FIRE MARSHAL AND/OR ADVISORY COUNCIL

MANDATORY SINGLE CODE

(APPLIES EVERYWHERE)

STATEWIDE MINIMUM STANDARD WITH LOCAL OPTION

ADOPT ONE OF EXISTING CODES AS MINIMUM

ADOPT SEVERAL EXISTING CODES AS MENU FOR LOCAL AREAS TO CHOOSE FROM

DRAFT OWN MINIMUM STANDARD CODE

ELEMENTS

1. PENALTIES OR INCENTIVES
2. TERM OF COMPLIANCE
3. RETROACTIVITY/ GRANDFATHERING
4. REQUIREMENT FOR PERIODIC UPDATING
5. EXEMPTED OCCUPANCIES
6. BUDGET CONSIDERATIONS
7. CONSOLIDATION OF ALL FIRE REGULATIONS UNDER A SINGLE STATE AGENCY

SOME FORM OF STATEWIDE STANDARDS

LOCAL STANDARD ONLY (OPTIONAL)

(STATUS QUO)

FIRE CODE

BUILDING CODE

(TREE WILL BE SIMILAR TO FIRE CODE)

Figure 2-1 Decision tree used by a Texas legislative committee in reviewing fire and building code options. (Developed by Tommie and Sindle Tomforde, Committee Staff.)

Building Officials, took over responsibility for its development and publication. Coordination of the Code with the Uniform Building code was a factor in this move.

In 1991 responsibility for development and publication of the Uniform Fire Code shifted to a new organization, the International Fire Code Institute, with offices in Austin, Texas. Voting membership is available to all members of the fire service. The code is used primarily in the western, Rocky Mountain, and Plains states.

A fire prevention code prepared by the Building Officials and Code Administrators (BOCA) International, Inc., is designed to be used in coordination with the BOCA National Building Code. Many communities using the Basic Building Code have adopted the BOCA Fire Prevention Code.

The Southern Building Code Congress International also produces a companion fire prevention code. This code is developed and endorsed by the Southeastern and Southwestern Associations of Fire Chiefs. Members of the Chiefs' associations must approve all changes in this code.

Model Building Codes

Building codes play an important part in the protection of the community from a fire safety standpoint. Code coverage includes basic structural requirements, structural integrity, fire protection as related to structural elements, means of egress, interior finish, vertical and horizontal openings, and many other areas that are directly related to fire protection. The building official must by necessity be concerned with all fire prevention and fire protection requirements with respect to original construction. Unless building officials and fire department personnel enjoy a close working relationship, possible conflicts of authority could be detrimental to community service and safety.

Model building codes sponsored by three national associations have gained national recognition and are widely used in communities throughout the United States. Most jurisdictions have adapted one of the nationally recognized model building codes to suit certain local needs, usually to establish more stringent requirements; other jurisdictions have developed their own local building codes. Several bills have been introduced in Congress to establish a federal building code for national application, but so far such a law has not been passed.

The three model building codes are described in the following paragraphs:

BOCA National Building Code

The BOCA National Building Code, widely used primarily in eastern and midwestern states, is published by the Building Officials and Code Administrators International, Inc. This organization, which has headquarters in Country Club Hills, Illinois, has published model building, plumbing, mechanical, fire prevention and property maintenance codes on a periodic basis since 1950. Membership is composed of state and local governments with representatives from all enforcement disciplines; however, memberships are available for all members of the construction community, including architects, engineers, and related professional groups. Members have the opportunity to present recommended changes in the codes for consideration and possible adoption. The organization holds an annual meeting at which code changes are discussed and final action is taken on adoption. The codes are updated and revised at three-year intervals. The BOCA National Codes were formerly known as BOCA Basic Codes.

Article 9 of the BOCA National Building Code is devoted entirely to fire protection, fire extinguishing equipment, and fire alarm systems. Other sections also relate to fire protection: Article 10 encompasses means of egress; article 6 includes fire resistance rating of structural elements and other matters relating to building size and construction; article 4 contains a number of fire-related subjects, including provisions for liquefied petroleum gases, paint spray operations, and motor fuel service stations; article 7 includes specific requirements on fire-resistive construction; and article 27 pertains to electrical wiring. These are just a few examples of the fire-related sections of the BOCA National Building Code. The other model codes described here cover similar subjects with respect to fire prevention and protection.[4]

Uniform Building Code

The Uniform Building Code is produced by the International Conference of Building Officials in Whittier, California. The Uniform Building Code was first published in 1927 and has become the primary building code in the Pacific Coast, Rocky Mountain and Plains states. This code, which is developed in very much the same manner as the BOCA and Standard codes, is updated on a periodic basis. Wood construction in conjunction with sprinkler protection is often recognized in lieu of construction having a greater fire resistance in the Uniform Building Code. Building officials comprise the voting membership of the conference.

Standard Building Code

Used primarily in the southeastern and southwestern states, the Standard Building Code was first published by the Southern Building

Code Congress International in 1945. This code is updated on a periodic basis and includes many items of special importance in the areas it serves. An example is the stringent coverage given to hurricane protection. From a fire protection standpoint, this code is organized in a manner somewhat similar to the other model codes. It is prepared in loose-leaf form so as to permit easy replacement of pages. The code is completely updated on a periodic basis, but immediate changes can be made on an annual basis through the use of the loose-leaf format. The Congress has its headquarters in Birmingham, Alabama.

Canadian Codes

Canadian Fire and Building Codes are developed and updated by committees of Canada's National Research Council. The codes are generally adopted at the provincial level with local enforcement.

National Fire Protection Association Standards _____

All three of the model building and fire prevention codes adopt selected National Fire Protection Association (NFPA) standards as portions of the code. Among the more popular standards adopted by reference are Flammable and Combustible Liquids Code (NFPA No. 30); Liquefied Petroleum Gas Code (NFPA No. 58); National Electrical Code (NFPA No. 70); and the Life Safety Code (NFPA No. 101). In a few cases there are conflicts between the specific provisions of the individual building code and the NFPA standard adopted. An example is the Life Safety Code (NFPA 101), in which travel distances to exitways and other specific requirements do not always coincide with building code requirements.

Board for Coordination of Model Codes

As noted in Chapter 2, considerable effort is made among the model building code groups and the National Fire Protection Association to standardize definitions and to eliminate conflicts and discrepancies such as the one mentioned previously in connection with the Life Safety Code. Periodic meetings are held, with representatives of the model code groups coming together to iron out conflicting requirements with the retention of individual requirements covering conditions peculiar to that of intended geographic location covered by the code. An example is the stringent earthquake coverage in the Uniform Building Code, which is used primarily in the West, as compared to the lesser degree of coverage in areas where such a natural phenomenon is less likely to occur. The Standard Code is known for its attention to hurricane protection.

The three model building code organizations in the United States are working to achieve common formatting among their codes. Although specific requirements will not necessarily be the same in each code, formatting will be the same, facilitating, for example, the use of multiple codes by architectural firms practicing in several regions. The American Institute of Architects and the Society of Fire Protection Engineers are assisting the Council of American Building Officials in this project.

Mini–Maxi Codes

The concept of the mini–maxi codes has caused a great deal of concern in fire service circles. Under this concept, municipalities and counties within a state adopting such a code are prohibited from adopting codes that are less stringent or more stringent than the state code. This approach gains uniformity of application but is seen by many as inhibiting a local jurisdiction from adopting more stringent fire protection requirements, such as stronger sprinkler provisions, which may be needed to address unique conditions.

Zoning Codes

Zoning codes also have an effect on fire prevention in a community. Through proper zoning, various types of occupancies are limited to given sections of the community. Bulk storage of flammable liquids would not, under a zoning code, be found in the middle of the high-value mercantile district. This condition does exist in a number of communities over the country; however, in practically all cases, flammable liquids storage was set up at the location before the advent of zoning or fire prevention regulations.

Zoning provisions also help in making sure that adequate clearances are provided across streets. This affects the potential for a conflagration in the community. Explosives storage and manufacturing are likewise generally very tightly controlled by local zoning requirements.

Electrical Codes

Electrical codes are primarily fire prevention codes in that electrical safety is so closely related to fire prevention. The principal purpose of electrical inspections is to see that wiring is safely installed in such a manner that people in the structure will not be directly endangered and that fires will not be started as a result of faulty installation.

There is a greater degree of uniformity of code usage in the electrical field than in any other field of public safety code coverage. The National Electrical Code, developed by the National Fire Protection Association, is incorporated as the electrical code in practically every state, as well as in practically all local jurisdictions having an electrical code in effect. Methods of enforcement vary; however, municipal electrical inspection is probably the most prevalent method employed.

In some areas, electrical code enforcement is at the state level, while in a few areas this responsibility is carried out by a private inspection organization. Work arrangements and procedures followed are essentially similar to those of any public electrical inspection agency.

Electrical inspections, like building code inspections (discussed in Chapter 5), are generally conducted on a fee basis, with the fee being sufficient to offset the cost of inspection. Normally the cost of the inspection service is passed on to the consumer in the electrical contractor's charges for the job. Building and electrical inspections have another feature in common—the necessity for close surveillance at the time of construction to achieve satisfactory code compliance. A delay of a day or two in making such an inspection can result in overlooking a serious condition from a safety standpoint. The enclosure that might have been completed during that delay might make proper inspection impossible.

As with building code inspection, there is a need for close coordination between fire prevention and electrical inspection agencies. Many fire prevention inspectors are not well versed in all details of the electrical code. It is well for them, therefore, to rely on the findings of the electrical inspection agencies in this technical field.

In most jurisdictions, enforcement of the electrical code is at the point of the provision of the electrical service. Most jurisdictions require full code compliance prior to the connection of service to a building from the public utility or power company. This procedure is quite effective and can also be used as an enforcement lever in the inspection of existing electrical installations. A threat of discontinuance of service is frequently more effective from an enforcement standpoint than a threat of a fine or other punitive measure.

A major problem in electrical code enforcement is the home wiring done by people untrained in the finer points of electrical work. The code enforcement agency has little control over such wiring, because normally a do-it-yourself electrician bypasses the formality of obtaining necessary permits.

Another factor in electrical code enforcement is the requirement for a high degree of technical competency, including licensure procedures, for electricians at various levels of expertise. This is necessary because of the nature of the work; much of the potentially dangerous wiring is hidden on completion of the structure. Only technically competent individuals should engage in this activity. The licensing boards for examination and certification of electricians are often closely related to the electrical code enforcement agencies.

Housing Codes

Another type of code that has a bearing on fire prevention is the housing code. In recent years, most larger communities in the United States have adopted housing codes in an effort to assure all citizens of adequate housing facilities.

Housing codes as a rule include a number of regulations relating to fire prevention and life safety. For example, means of egress and heating appliances are covered by code provisions. Space heaters are not permitted under model housing codes. The right of entry for inspection under housing code authority is generally broader than that for fire prevention inspectors. This provides the housing inspector with an instrument for upgrading life safety in many occupancies that may be inspected by fire department personnel only on invitation.

Because of the personal nature of housing code enforcement, many jurisdictions have been reluctant to permit the implementation of such a code. Many citizens feel that obligatory inspections of individual homes under a housing code represent an unwarranted intrusion on privacy.

It is highly desirable for people having a fire prevention responsibility in a community to maintain close liaison with those responsible for housing code enforcement. This might include having fire safety personnel speak at housing code enforcement training sessions, as well as other contacts.

Plumbing, Heating, and Air-Conditioning Codes

Mechanical codes, including heating, ventilation, and air-conditioning as well as plumbing codes, will be treated together because they are usually enforced by the same agency in a community. Plumbing codes have a bearing on fire protection, especially when requirements related to automatic sprinkler protection are included as a part of the plumbing code. The plumbing inspector may be responsible for seeing that automatic sprinkler equipment is properly installed. The inspector may also have responsibility for checking underground connections, relating to fire protection services and for checking hydrants, standpipes, and other features relating to fire protection water supplies.

Mechanical codes, which cover heating and air-conditioning, have a major effect on fire prevention in that they usually include coverage on duct work and other matters related to distribution through heating and air-conditioning systems. Improper installation has resulted in the spread of smoke and fire; in fact, this sometimes occurs even with proper installation. It is highly important that heating and air-conditioning installations be made by individuals who are thoroughly conversant in modern fire protection practices. Dampers, floor penetrations, smoke detectors, and duct materials are among fire safety considerations in these installations.

Other codes having a bearing on structural fire safety include codes pertaining to elevators; elimination of architectural barriers to the physically handicapped; energy conservation; gas fitting; historic preservation; manufactured buildings; mobile homes; and pressure vessels and boilers. Many states and communities have not adopted all of these codes. Where adopted, responsibility for enforcement is varied.

Codes for elimination of architectural barriers relate to exit facilities, fire alarms, and other fire safety features. Energy conservation codes refer to construction materials and other fire safety concerns. Gas codes, which regulate gas piping and installation, have a major impact on fire safety, as have elevator codes and pressure vessel and boiler codes.

Historic preservation codes are enacted to encourage the retention of historically significant structures in a community. These codes may permit less stringent fire protection requirements for existing buildings than for new ones. It is impossible to renovate some historic buildings in such a manner as to be in full compliance with codes for new structures.

Forestry Codes

Usually administered at the state level, forestry codes and regulations have a definite effect on fire prevention. These regulations are generally in effect only within a given number of feet of wooded or forested areas, and in many states they are not applicable within incorporated municipalities. Regulations often prohibit open burning during periods of low humidity. They may also require safety precautions in connection with the use of matches and smoking materials.

Forestry laws are generally comprehensive and prohibit a wide variety of unsafe acts in forest lands. Among acts generally prohibited are leaving campfires unattended, open burning without a permit during certain seasons, operating a vehicle in forest lands without a muffler, and dropping or throwing burning matches, lighted cigarettes, or other burning materials in or near woodlands. The operation of railroads in woodlands is also closely regulated because of the danger of fires as a result of sparks. Adjacent lands are required to be cleared to reduce this danger.

In many states people who set fires on property they own and permit the fires to escape and become uncontrolled are responsible for payment of all costs for fighting the forest fires. The fires may have been set originally to destroy debris or rubbish.

Forestry regulations are usually very well-known, at least by people who work in the forested lands and by those who frequent them for recreational purposes. Because of the large landmasses covered by such regulations, enforcement is often difficult and may be compared to enforcement of maximum speed limits on major interstate highways. It is most uncommon for an enforcement official to see a violation being committed, and enforcement generally comes about through individual complaints with occasional spotting of dangerous practices during routine patrol.

Environmental Regulations

Air pollution regulations, which are normally enforced by health and environmental agencies, have a bearing on fire safety. Regulations usually require permits for all open burnings; they also impose severe restrictions on locations at which open burning may be carried out and specify narrow time frames for necessary burning.

These regulations have a positive effect in the fire prevention field through the requirement of burning permits and consequent patrol activities designed to reduce nonapproved burnings. Air pollution regulations may preclude the burning of buildings by local fire departments for hazard abatement and training purposes. Normally, some provision is made to permit bona fide training activities to take place; however, this is not always the case.

Another way in which environmental regulations may conflict with fire prevention is by creating the need for storage of combustibles awaiting pickup. Wastepaper and other trash that would normally be burned is held for a period of time awaiting pickup by trash trucks for delivery to the proper location. During this storage time the hazard within the facility may be increased as a result of holding these combustibles.

Again, as in the case of housing code enforcement, it is highly desirable for the fire prevention inspector to maintain close liaison with personnel in the air quality control field. Many misunderstandings can be resolved by this cooperation, particularly with respect to requirements for burning permits. A permit from air quality control is often required in addition to the fire department permit in communities that operate under a model fire prevention code. Forestry agencies may also require permits, so it is conceivable that a person might have to get three different permits just to burn one small pile of trash in the yard. To eliminate this kind of overlap, some communities have combined air quality control, forestry, and fire prevention bureau permit requirements.

Public support for air quality control measures has been strong, which works to the advantage of efforts to achieve 100 percent control through issuance of permits. A fairly high percentage of burnings are being reported to the air quality control people because of public interest and support. Many of these calls are in the form of complaints by neighbors who take exception to burnings occurring in their neighborhood.

Environmental protection regulations enforced by the federal government, and by states and local jurisdictions, have important implications in fire prevention. These codes may encompass air quality control, and they may include features relating to other environmental problems, namely those at the ground level. From a fire protection standpoint, these codes have a bearing on disposal of dangerous chemicals, for example. Regulations might prohibit disposal of such chemicals and impose rigid requirements for dumping operations. Such requirements could be important to fire safety in the community, especially if flammable liquids were involved.

Although often included in an environmental protection regulation, water control may be treated as a separate public safety function. Water resources are, of course, vital to any community, and regulations for safeguarding this resource are most essential. Water resource protection regulations would have an effect on fire prevention in that oil pollution is almost always covered. There should also be joint efforts between water resources and fire service personnel in the handling of leaks, spills, and other incidents where waterways or water sources may be contaminated.

Other Areas of Regulation

Miscellaneous regulations that have some connection to fire prevention are being introduced. Some jurisdictions have enacted requirements for licensing of television and appliance repair technicians. This has a definite bearing on fire prevention in that higher quality repair should reduce the number of fires caused by faulty appliances. The jurisdiction can revoke the individual permit in the event of poor performance.

The complexity of urban life today has brought about a profusion of regulations. Public demands for higher standards of performance on the part of repair personnel will no doubt bring about future licensure requirements in a greater number of service businesses.

Fire Safety Considerations in Special Occupancies _____

Fire prevention and life safety measures in health care facilities and high-rise structures have all too often come about only when public concern is aroused in the aftermath of tragic fires. These areas are receiving more attention of late at all levels of government.

Nursing Homes

An example of the evolution of fire prevention and protection requirements pertaining to a specific type of occupancy is that of nursing home fire safety. Because the potential victims are generally physically or mentally incapable of fending for themselves in the event of fire, the relative safety of nursing homes is based almost entirely on the effectiveness of code requirements.

The nursing home field is a developing one because of the increase in average life expectancy. There also seems to be a growing reluctance by families to keep elderly relatives in their homes when they can no longer actively participate in daily family life. Programs such as Medicaid and Medicare that provide financial assistance to the elderly have likewise encouraged the development of nursing homes and other facilities for the extended care of the aged. The Medicaid program is primarily designed to assist those who are indigent, while the Medicare program provides assistance to those who have participated in the Social Security program through their work during more productive years. All of these factors have combined to create a great demand for nursing home care and other provisions for housing elderly people.

As the industry expanded, little thought was given in some localities to the fire safety of the structures. In many cases old, large dwellings that were no longer suitable for family occupancy were converted into nursing homes. The average nursing home years ago was of this type. Few structures were built specifically for the purpose of providing such care. In some jurisdictions, fire personnel equipped with adequate codes required the provision of automatic sprinkler protection and other safeguards in these facilities. In some cases strict requirements, such as sprinkler protection, enclosure of vertical openings, and other safeguards, caused nursing home operators to locate their facilities outside of the city limits in an area where regulations were not so stringent.

A number of fires that resulted in multiple fatalities occurred in substandard structures in practically every section of the United States. In some states, these fires brought about modifications in fire prevention regulations, including requirements for automatic sprinkler protection and other fire safety features for all other nursing homes. Several jurisdictions found, only after a major nursing home fire, that there was no agency in the community or state that had responsibility for fire safety enforcement in such facilities.

The interest of the federal government in fire safety increased because of the previously mentioned programs for federal financial assistance to many elderly citizens. The United States Congress, as well as the Department of Health, Education and Welfare (now Health and Human Services) recognized in the late 1960s that many of the patients receiving federal government subsidies were not being housed in safe environments. A number of states were not ensuring a maximum degree of safety for these people. There was also disagreement as to what were the best methods for providing adequate safety. Some believed that the use of fire detectors employing heat-sensing devices was the best. In several states, heat detectors were required in institutions housing elderly people. In other states, authorities felt that smoke-detection equipment throughout the structure would provide an adequate degree of protection. Automatic sprinkler systems were considered in some states as the only adequate means of affording fire protection.

For all the concerned citizens, including fire officials, who thought that improved fire safety was needed, there were others who thought that economic considerations would not permit compliance with more stringent fire safety requirements. Some felt that stringent requirements would cause many homes to close and that the residents would then have no place to reside. The Department of Health, Education and Welfare attempted to put into effect extensive fire safety requirements, including mandatory installation of automatic sprinklers. These attempts were thwarted for many months by people who claimed that economic factors would preclude effective implementation of the requirements. Fires resulting in multiple fatalities continued to occur, and the pressures to do something became greater and greater. Eventually, federal regulations went into effect in 1971 that required automatic sprinkler protection and other fire protection safeguards in many nursing homes. These regulations have had great impact in jurisdictions that had not previously required such protection. As could be expected, many of the same appeals were heard again, and it was necessary to discontinue funding in a few cases to enforce compliance.

At the present time the federal program is in full effect for skilled and intermediate-care facilities. Enforcement of the regulations is federally funded and carried out by state personnel, normally through the state fire marshal's office or state health department.

Boarding Homes for the Elderly

Adequate safeguards for elderly residents of board-and-care homes, or boarding homes, as they are generally called, are far from universally provided. Many elderly people are housed in old buildings of frame or ordinary construction, and very few are provided with automatic sprinkler protection or any other meaningful protective devices. A U.S. House of Representatives' report on health care facilities, mentioned below, made recommendations with respect to this problem, but progress in adequate safeguards will probably be limited until public interest is sufficiently aroused. The interest shown by Congress may serve to hasten action, but it is unfortunate that additional lives will probably be lost in these facilities before adequate safeguards are required in all jurisdictions. The problem is aggravated because control at the state level for facilities of this type is somewhat limited.

The U.S. Congress authorized a special study of health care facilities as a project of the Committee on Government Operations of the House of Representatives, which was carried out in 1972. The committee's report strongly recommended increased life safety for all health care facilities, including extended-care facilities and so-called boarding homes that house primarily elderly people.[5] The committee recommended the continued use of federal financing as a lever to improve fire safety in health care facilities.

High-Rise Structures

A good example of how standards, procedures, and code requirements for fire prevention and fire protection evolve is the recent history of high-rise structures. Interest was focused in this direction as a result of multiple-fatality fires in high-rise structures that occurred in Atlanta, Georgia, and New Orleans, Louisiana, in the winter of 1972. The New Orleans structure was used for office occupancy with some public assembly use; the Atlanta structure was an apartment building for the elderly. Both structures were quite modern, and construction was generally fire-resistive.

The fire in New Orleans gained national attention because of nationwide live television coverage. Through this exposure, a great many people became aware of a fire safety problem in high-rise structures that before the advent of television might have generated only local concern. The Atlanta fire called attention to problems in housing for elderly people who are capable of living independently.

Both the Atlanta and New Orleans fires resulted in the immediate drafting of legislation, municipal ordinances, and regulations providing for fire protection in high-rise structures in a number of jurisdictions. In Maryland, for example, the General Assembly in 1974 enacted legislation requiring automatic sprinkler protection in all new high-rise buildings constructed in the state.

Louisiana, Massachusetts, and Nevada have enacted legislation requiring automatic sprinklers in existing as well as new high-rise buildings. Agricultural buildings are generally exempt from the provisions. Codes in some communities have been modified to require retroactive installation of sprinklers in high-rise buildings. Atlanta, Denver, Philadelphia, and San Francisco are examples of this change.

In some cases tragic events have led to rapid enactment of fire safety codes. This has been part of the history of code development, as noted in Chapter 1. An example of this phenomena was the legislation enacted in the state of Nevada after the MGM Grand Hotel and Hilton Hotel fires in 1979. These tragic events led to the formation of that state's Governor's Commission on Fire Safety Codes. This commission studied fire safety and building codes in Nevada.

The commission met a number of times and reviewed pertinent codes. It also took testimony from fire safety experts and subsequently recommended enactment of a strong fire code that required automatic sprinkler protection retroactively in many Nevada buildings. Citizen pressure generated by these fires helped the commission get its findings rapidly adopted, resulting in major improvements in fire safety in publicly occupied buildings.

A similar chain of events occurred in the province of Ontario after a fire took six lives in the Inn on the Park in suburban Toronto in May 1981. The provincial government's exhaustive study of the fire led to a number of code changes.

Florida enacted a law in 1993 requiring full automatic sprinkler protection in all new buildings three or more stories in height. One- and two-family dwellings are not included in this requirement.

Residential Sprinklers

Another example of the code development process is that relating to residential automatic sprinklers. The development of mandatory requirements for residential sprinklers has taken a long and tortuous route.

While the value of automatic sprinklers has been recognized for over 100 years in industrial, commercial, and institutional occupancies, little credence has been given to the application of these principles to residential occupancies. The one exception was the use of automatic sprinklers in hotels and large apartment houses.

In the late 1970s the city of San Clemente, California, pioneered requiring automatic sprinklers in residences located beyond usual access limitations for fire apparatus. The city declared that residences built more than five miles from the nearest fire station had to be equipped with residential automatic sprinklers. The organizations developing the standards and the sprinkler manufacturers had to adjust to this new demand. The sprinkler industry began producing suitable equipment on a limited basis while a standard was developed and published.

The concept that "a man's home is his castle" has made this type of regulation difficult to implement. In addition, many home builders' groups have strongly opposed any measure that would bring about an increase in the construction cost of new dwellings. Home builders feel that even a minor increase in cost will reduce the number of people who are able to buy new homes.

A number of jurisdictions have adopted regulations similar to those in San Clemente. In fact, San Clemente subsequently changed its regulation to require sprinklers in all new residences, a feature that is encompassed in most of the other residential sprinkler ordinances.

In 1985 Scottsdale, Arizona, became the largest city in the United States (with a population of over 148,000) to require residential sprinklers in all new residences. The city passed a requirement in 1974 requiring sprinklers in all new mercantile, industrial, and commercial structures with more than 7,500 square feet on the first floor or with two or more stories in commercial buildings or three or more stories in hotels or apartment complexes. This requirement has been so successful that efforts to convince elected officials as to the need and desirability of residential sprinkler requirements were made much easier. A number of cities in California, including Salinas, are requiring automatic sprinklers in all new residences. Such a requirement has been imposed in Dover, New Hampshire, Prince George's County, Maryland, and many other jurisdictions.

Cobb County, Georgia, has also been a leader in residential sprinkler installation. Cobb County uses trade-offs rather than a specific ordinance as the method for securing sprinkler installations. The building official, with fire department support, has granted a number of trade-offs in multifamily residential construction requirements that result in such buildings being constructed less expensively with sprinklers than without. For example, two-hour fire rating separation requirements are reduced to one hour when the structure is sprinklered, resulting in a cost saving. Practically all contractors building multifamily residences in Cobb County are using this alternate means of construction, thereby effecting a savings in construction cost. Of course, the life expectancy of the building and its occupants from a fire safety standpoint is much greater, and the apartment units are more marketable.

The use of residential sprinklers is expanding and at least one state, North Carolina, has changed the state building code to recognize residential sprinkler capabilities in the same fashion as in Cobb County. North Carolina has a mini–maxi state building code, so state recognition is essential for local jurisdictions to be able to permit such trade-offs.

California's League of Cities issued a fire sprinkler informational packet to all mayors and city managers in which installation of these systems is strongly supported. The League also passed a resolution urging cities to adopt ordinances requiring such installations in all new structures, including single family residences.

Canadian cities are also requiring sprinkler systems in new construction, including residences. Westmount, Quebec, imposed this requirement in 1989, as did Quebec City. Vancouver, British Columbia, later became the first major North American city to require sprinklers in all new buildings.

An example of code development was the successful effort by the Florida Fire Chiefs Association and other organizations in that state to secure legislation that retroactively requires automatic sprinklers in existing hotels, motels, and time-share occupancies. This legislation was promoted by the fire chiefs with the support of the hotel and motel associations, League of Cities, and other interested groups. It came about without a multi-fatality fire as an impetus; however, the sponsoring agencies recognized the value of the tourism industry to the state and the devastating effect that such a fire could have on business. Previous attempts by the sprinkler industry to get this type of legislation enacted had failed, partly because they were looked upon as self-serving. A concentrated effort, plus a reasonable time period for compliance, brought about enactment in 1983.

Another consideration in the development of codes is that of evaluating new conditions, innovations, occupancies, and processes. As an example, the advent of the "hotel-hospital" brings a challenge from a code application standpoint. These facilities are designed for individuals who no longer require close hospital care but have recovered to the extent that they can exist in a homier, less expensive setting near the hospital. The latter feature is considered important because of the possibility of a relapse. The code developer and enforcer is concerned with appropriate fire safety measures for this new type of occupancy. Is it considered "institutional," or is it considered "residential"?

Another example is that of the development of the semiconductor industry. When faced with hazards of these occupancies, which were first developed in California, the western fire service was quick to develop a section of the Uniform Fire Code relating specifically to the semiconductor field. One problem in such a situation is that little empirical data are available to substantiate code requirements.

Summary

Various types of codes in effect in the nation have a bearing on fire prevention. Most of these codes are enacted and enforced at the municipal level, some are at the state level, and a few are national.

Fire prevention codes are usually the province of the fire department and are principally concerned with fire safety regulations in the maintenance and use of buildings and premises and with respect to hazardous materials, processes, and machinery used in buildings. Building codes, which are usually enforced by the building department, are also concerned with fire safety regulations, but with respect to construction, location, and use of buildings.

The fire prevention code as we know it today can be traced to the influence of the insurance industry, which has played an important part in public fire protection in this country. Following the great Baltimore fire of 1904, the National Board of Fire Underwriters (whose name was changed in 1965 to American Insurance Association) developed a Standard Grading Schedule for cities and towns with reference to fire defenses, which gave great impetus to the development and enactment of fire prevention codes. Model building codes and fire prevention codes that have contributed greatly are produced by three national associations: Building Officials and Code Administrators International, Inc., International Conference of Building Officials, and Southern Building Code Congress. All nationally recognized model building codes incorporate selected standards of the National Fire Protection Association, which has made significant contributions in developments for the abatement and prevention of fire.

Fire prevention and life safety measures in special occupancies, such as health care facilities, schools, and high-rise structures, have all too often come about only when public concern is aroused in the aftermath of tragic fires. These areas are receiving more attention of late, with the advent of federal regulations with respect to nursing homes for the elderly.

In summary, the development and enactment of codes for the abatement and prevention of fires should always be carried out with the following quotation in mind: "The laws of our land build on reason and experience—and the chief of all laws is the will of the people." This quotation from the Mobile County courthouse in Mobile, Alabama, expresses the need for the consideration of factors that should never be overlooked in the development and promulgation of regulations. Certainly, enforcement cannot be adequate if any of the mentioned principles are violated.

Notes

1. *Fire Suppression Rating Schedule* (New York: Insurance Services Office, 1980).
2. *Fire Prevention Code* (New York: American Insurance Association, 1976).
3. *Fire Prevention Code NFPA No. 1* (Quincy, Mass: National Fire Protection Association, 1992).

4. *BOCA National Building Code* (Country Club Hills, Ill.: Building Officials and Code Administrators International, Inc., 1993).

5. U.S. Congress, House, Sixteenth Report by the Committee on Government Operations, *Saving Lives in Nursing Homes* (Washington, D.C.: U.S. Government Printing Office, 1972).

CHAPTER 3

Organization and Administration of Municipal Fire Prevention Efforts

Fire prevention has been recognized as a function of municipal government since the earliest days of this country, as discussed in Chapter 1. In the colonial period, the administrative functions of the town with respect to fire, police, sewage, water, health, and street lighting were often carried out through volunteer citizen service. Some of the early municipal charters required citizens to serve in certain positions of responsibility without pay, and people usually readily accepted such responsibilities. Voluntary fire protection and fire prevention service is provided in thousands of communities in the United States and Canada.

The administration of municipal government today encompasses a wide variety of functions, most of which have direct and far-reaching impact on all citizens. Taxation to support public services is universal; for this reason, if for no other, the citizen taxpayer is concerned about the fire protection and fire prevention service in the community. Many county governments provide services similar to those of municipalities.

The Place of Fire Prevention in Municipal Government

By tradition, fire prevention and fire protection in the United States and Canada have been primarily local government responsibilities. Fire and fire-related problems are more locally confined than are most other concerns of public services, such as crime, health, welfare, and employment. It is quite unusual, for example, for a fire to actually burn in more than one jurisdiction. Furthermore, fire safety code enforcement can usually be most effectively carried out at the local level.

In most communities a fire department is usually formed when something happens to demonstrate the need for fire protection for public safety and welfare. A community typically gets along without a fire department until it becomes obvious to the citizens that they can no longer rely on neighboring communities to provide fire protection or until a large fire involving loss of life or serious property damage brings urgent demands for community action to organize a fire department.

In most areas of the United States and Canada, the incorporated municipality is looked to as the most logical agency to develop and operate a fire department. In some areas, fire protection is not considered a municipal function, and service is provided by an incorporated fire department. This arrangement is often used in unincorporated communities.

In municipal government, elected officials are responsible for the overall administration of government. Administrative heads of departments, who are responsible for carrying out the day-to-day functions of the government, may be political appointees or may be appointed through a merit system. Lower ranks of municipal employees are usually selected through merit system or civil service procedures.

Recognition of the close interrelation of municipal functions is essential for good city government. The lack of coordination among municipal functionaries may be a reflection of some weakness on the part of the city administration, or it may be that the individual departments are more concerned about preserving their autonomy than performing as part of a city government team. Some fire department administrative personnel have a tendency to believe that fire protection is so specialized that it has little relationship to other municipal government functions. The fire department is an important entity in municipal government, but to function most effectively it should cooperate with other municipal agencies in the mutual aim of improved public service.

Charter Responsibilities

The charters of most cities include fire protection as a responsibility of the municipality. The provisions of some municipal and fire department charters may make specific mention of fire prevention, although they generally encompass overall fire protection, which may easily be interpreted to include fire prevention. An example of the former is the charter of the College Park, Maryland, Fire Department, which states the following purpose for existence: "Its object shall be the prevention and extinguishment of fires and the protection of life and property in College Park and vicinity."

Even if fire is not specifically mentioned, the responsibility for public safety is clearly drawn in the charters of practically all cities. There is no question that the public safety includes protection from fire, so it may be assumed that broad public safety provisions give charter backing for appropriate actions and measures in fire prevention.

Fire Service Administration in Municipal Government

With thousands of incorporated cities and municipalities in the United States and Canada, it is understandable that there is a wide diversity of public fire services, ranging from volunteer or call groups in smaller communities, combination volunteer and career forces, to fire departments with specialized divisions for fire prevention. In the traditional municipal fire department, the fire chief reports directly to the mayor or other top administrator, an arrangement that makes fire safety responsibility equal to other municipal functions.

In another form of organizational structure found in some communities, the fire department is placed under a director of public safety, who also has the responsibility of overseeing the operation of the police department, the jail, and possibly building inspections and weights and measures functions. Division heads report to the director of public safety. This arrangement places a high percentage of municipal employees under one agency head; in fact, the number sometimes exceeds the sum of all other municipal employees.

Another variation is the combination police–fire department concept employed in some jurisdictions. This type of organization has been advocated as a means of saving money by reducing personnel costs, but it has not met with wide acceptance. Communities in Michigan, California, Florida, and other states have joint police and fire agencies, and a number of communities in the province of Quebec follow this pattern. Other communities have adopted the idea on a limited basis; that is, only a few stations within a fire department are set up to operate as combined police–fire stations, or certain functions such as personnel and purchasing are combined.

Fire Prevention Functions

Among the stated objectives of a fire department, the prevention of fire is a primary goal. As mentioned earlier in this chapter, fire prevention is considered as an important component in the fire defenses of a community. Fire prevention refers primarily to measures directed toward avoiding the inception of fire. Among these measures are included fire safety education, fire prevention inspection, fire code enforcement, investigation of fires to determine causes, and investigation of suspicious fires. Analysis of information gained from the latter two functions can be of great value in evaluating and improving fire prevention practices to help prevent fires.

There are many organizational approaches to assigning responsibility for fire prevention functions. Some fire departments do not consider fire prevention a full-time activity and may assign fire prevention responsibilities as collateral duties.

In some cities, the fire prevention bureau or fire marshal's office is a separate arm of government, with the fire prevention chief or fire marshal reporting directly to the city manager or chief executive. In some fire departments, fire prevention responsibilities are considered to be line functions for suppression forces. A number of federal fire departments, as well as industrial fire brigades, are organized with fire prevention functions as a line responsibility for suppression personnel. These departments make sure that personnel are advised of such requirements on entry to the service. They are employed not only as fire fighters but as fire "preventionists" as well.

The Fire Prevention Bureau _____

In fire departments serving communities with populations of 25,000 or more, the most common organizational concept is a fire prevention bureau that functions as a separate arm of the fire department. Personnel are assigned to the bureau on a full-time basis and normally have no fire suppression responsibilities, except that in emergencies they may be pressed into service in fire fighting. In departments where there is a separate fire investigation bureau, fire safety personnel may on occasion be called on to assist in investigations.

The fire prevention bureau coordinates all fire prevention activities within the fire department and makes important contributions to fire department public relations through its inspection and fire safety education programs. Fire prevention personnel are in direct personal contact with the citizens of the community.

In some fire departments, the attitude prevails that fire prevention work is for those who are physically unable to carry on active fire fighting duties. People who have been injured in the line of duty or who are suffering from some other physical disability that precludes their effective service in fire suppression are assigned to the fire safety bureau. This practice may be the result of a lack of understanding of the true role of fire prevention in the fire defenses of a community. Under the fire laws and regulations in effect in most jurisdictions, the fire department has the authority and obligation to carry out essential fire prevention measures. Furthermore, fire prevention work has some strenuous aspects; conducting inspections, for example, may require climbing steps or other physical effort that may be beyond the capacity of the physically handicapped.

There is no doubt, however, that some excellent people have been assigned to fire prevention bureaus because of physical disabilities that precluded their functioning in other sections of the department. In one example, the fire prevention chief was assigned to that duty because of a heart condition. Had it not been for this disability, he—a university graduate—would probably have become the chief of the fire department. As chief of the fire prevention bureau, he functioned in an outstanding manner and continued year after year in this capacity.

The One-Member Bureau

The fire prevention bureau should be organized in such a manner as to permit clear lines of control. In a smaller community only one individual may be assigned to the bureau, which makes the matter of organization simple. The one-person fire prevention bureau can be quite effective. Arrangements should be made to allow this person to take care of duties in the field by assigning individuals to take care of telephone calls in absences from the office and to take over during vacations or periods of illness.

Many smaller communities have outstanding fire prevention bureaus. Great care should be given to the selection of an individual to head a program of this type because of the range of duties required and because so much of the work involves contact with the public. After some years on the job and especially after becoming proficient at it, the lone fire prevention officer may become entrenched and possibly side-tracked from promotion because of proficiency in the field. Another problem in the one-person bureau is that the officer, because of constant public contact, may become better known in the community than the chief of the fire department, which could create internal problems. Close coordination and understanding between the fire prevention officer and the chief of the department are needed to maintain harmony.

Generalists versus Specialists in the Fire Prevention Bureau

In the fire department that is large enough to establish a fire prevention bureau consisting of more than one individual, it is important that the organizational concepts employed in the bureau be well thought out. There is a tendency to assign personnel to the bureau to get the job done, without much thought to appropriate individual assignments.

In a smaller fire prevention bureau, consisting of from two to seven people, it seems most practical to assign all personnel as generalists.

This would mean that any assignment for inspection, plan, review, fire prevention education, or investigation could be handled by any member of the bureau. Assignments might be made on a geographic basis. Personnel assigned to one area of town are responsible for all inspections within that area, regardless of the type of occupancy. This arrangement has the advantage of providing inspections at the most reasonable cost, because of less travel involved. The inspector can go to the assigned section of the town, get out of the car, and go from door to door, conducting inspections for an extended period of time. This would be the case in the business district or in shopping centers located in the outskirts of the town. One disadvantage of this system is the possibility that having inspectors working different parts of town could bring about variations in inspection procedures or in interpretations of code requirements for similar facilities. A nursing home in the northwestern part of town, for example, might not be inspected in the same manner as one in the southeastern part of town.

On the other hand, some fire prevention bureaus, especially the larger ones, find it advantageous to develop and assign inspectors as specialists in a particular field. One inspector would specialize in flammable-liquids installations, for example, another in health facilities, while another might handle all industrial occupancies. If the inspection load in a community does not warrant full-time assignments in the inspector's area of expertise, the specialist may spend some time in handling general assignment work. The disadvantages of the specialist system of inspections are that transportation costs are higher when inspections are not confined to one area and that inspectors have less opportunity to establish close contacts in the community than when working in one neighborhood. The advantages and disadvantages to the community and to the department should be thoroughly weighed to determine the best system for a particular organization.

Here, too, as with the one-person bureau, care must be taken to avoid dead-ending the specialist inspector because of expertise in a given field. Personnel assignment must take into consideration the promotional policies of the bureau and the fire department. Promotion to captain, for example, often requires several years' experience in fire suppression. The individual with a specialized background should not be penalized in movement within the department.

Suitable Quarters

In the development and organization of a fire prevention bureau, it is important to provide suitable quarters for adequate operation and public contact. Unfortunately, some fire prevention bureaus are tucked away in remote corners of the central fire station that are poorly located for public access. Often there are no facilities for public parking in the vicinity of the fire headquarters. In some cities the central fire station is located in a neighborhood that has become undesirable from the standpoint of safety on the streets.

The location of the fire prevention bureau within the fire station may be a source of internal problems. In some cases it is necessary for the fire prevention bureau "customer" to go through sleeping quarters or day rooms occupied by fire fighting personnel to reach the fire prevention bureau. This is a bad situation, which not only may cause inconvenience for the public, but may also result in resentment toward the fire prevention bureau on the part of the fire suppression personnel.

The physical location of the fire prevention bureau in the fire station or city hall may also be a factor in the public's attitude toward the fire prevention program. If the bureau is poorly located, the implication is that the bureau is of limited importance in the fire department and, in fact, in the overall municipal government. The fire prevention bureau should be located in a section of the fire station or city hall that can be easily reached by the public. Parking facilities should be made available for visitors.

Communication with the Public

The bureau offices should have adequate facilities for receiving telephone calls. Citizens calling to transact business with the fire prevention bureau should be able to reach a person who can supply the needed information without having to talk with several other people who may not be directly concerned with fire prevention. The fire prevention bureau should also have a clearly established mailing address.

Because of public uncertainty, the chief of the fire prevention bureau, or the fire marshal as some may be called, may receive communications, telephone calls, or visits that should properly have gone to the chief of the fire department. On the other hand, calls may be referred to the fire chief that should have been referred to the fire prevention bureau. With responsibilities for the overall administration and day-to-day operations of the fire department, the fire chief is not likely to be fully informed of routine fire prevention bureau activities. Perhaps some internal fire department method of screening phone calls and visits would help clear up confusions.

Working Hours

Personnel in the fire prevention bureau are often assigned shorter working hours than other fire department employees. There are many exceptions to this rule, however, as fire suppression personnel are given shorter workweeks. At one time, assignment to the fire prevention bureau might have been considered desirable from the standpoint of working hours, but this view has all but disappeared with changes in working conditions for fire suppression personnel.

Practice dictates that the fire prevention personnel work on a schedule comparable to the headquarters office personnel and to the city government office forces. The workweek is normally 40 hours, although some cities have reduced this to 37 hours or less. Some fire prevention bureaus work four 10-hour days per week. In a number of cities fire fighting personnel are now working 42, 48, or 56 hours a week. In some the workweek is down to 40 hours. Shifts are generally arranged on a 10-hour, 14-hour, or 24-hour schedule. In either case, employees have daytime hours off during the week, which allows them to participate in other activities during the normal daytime workweek. Such an arrangement cannot be made by an individual assigned to a 40-hour workweek in the fire prevention bureau. For this reason some cities are finding it difficult to encourage people to voluntarily enter fire prevention work. As a result, some cities are finding it necessary to upgrade fire prevention personnel and to take other steps to make the job more attractive.

Sources of Conflict within the Fire Department _____

One of the major responsibilities facing the fire chief is that of bringing about a full understanding and appreciation of the fire prevention bureau on the part of other fire department personnel. Misunderstandings are not limited to those connected with working hours. Many other areas of misunderstanding arise in connection with the operation of the bureau. Friction may develop between bureau and other department personnel because of a feeling on the part of fire suppression personnel that the fire prevention people do not have to take the physical punishment associated with fire fighting and emergency medical services.

A conflict may also arise between the two segments because of the procedure followed in many communities of assigning automobiles to fire prevention bureau personnel. These vehicles are assigned for official use only, including investigation of overcrowding complaints and inspection of nightclubs. However, a certain status symbol may be attached to possession of a city car, especially if it may be taken home at night. There are many justifiable reasons for assigning automobiles to fire prevention bureau personnel. It may be necessary for these individuals to respond to fires to carry out fire investigation duties.

Fire suppression personnel may resent the fact that fire prevention bureau personnel are in the public eye more frequently. Inspectors are in daily contact with business owners, educators, operators of institutions, and other community leaders in the normal course of their duties. Fire suppression personnel have little public contact except on the fireground and on emergency medical calls, as well as in their personal spheres.

Fire Prevention Advisory Committee _____

Another concept that has been found to be quite helpful in creating a better understanding of the role of fire safety is the fire prevention advisory committee, made up of members from all divisions of the fire department. One or two representatives might be selected from each of the battalions or districts within the department. The committee formulates the fire prevention program for the department and meets from time to time to evaluate their efforts and suggest changes in direction. Through the establishment of a fire prevention advisory committee, personnel throughout the department can be given an appreciation of fire prevention duties and responsibilities.

As an effort in the same direction, some municipalities appoint people from outside of the fire department as members of their fire prevention committees. These may be representatives of civic organizations within the community, or they may be selected from the public at large. This arrangement results in greater understanding of fire prevention by the public.

Mobile Fire Safety Team _____

Another innovation is the mobile safety team, which makes regular patrols to inspect structures and to take the message of fire prevention to the citizens of the community. The vehicle used for this purpose may be a carryall or a station wagon, which is usually outfitted with minimal fire fighting equipment and carries a supply of fire safety literature. Personnel assigned to such a unit are knowledgeable in both fire suppression and fire prevention. The mobile safety team may stop at a shopping center to conduct demonstrations and pass out literature; on occasion, they have actually discovered fires on routine patrols and have used the equipment in the mobile unit to fight the fire.

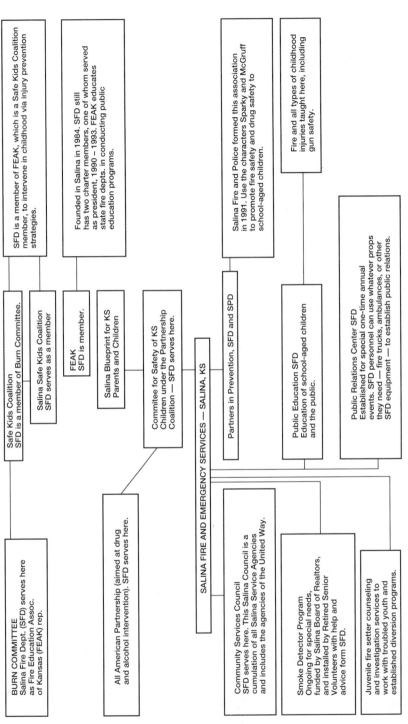

Figure 3-1 Example of comprehensive communitywide fire prevention/public safety program. (Salina, Kansas Fire Department.)

Company Fire Management Areas _____

One of the most promising concepts is that of the fire management area program for utilizing company personnel for fire safety duties in their response territories. This concept is described by Springfield, Illinois, fire department officials as follows:

Springfield has twelve fire stations. On the basis of still alarm districts, or at least approximately so, each station was given a geographical area, referred to as a Fire Management Area. This area was subdivided, assigning each shift one of the Sub-Fire Management Areas. Now we have 33 Sub-Fire Management Areas.

The company is responsible for the following items in each area, by shift: a) hydrant inspections, once annually; b) target hazard identification and pre-fire planning to include area fire flow tests. Each Sub-Management Area team was required to identify the most serious life hazard, property hazard, and key hazard (hazard if property loss would affect the community the most—power plant, hospital, factory, etc.) Each company studied their sub-management area and picked the three targets. They then were assigned a pre-fire planning inspection to include the aforementioned area fire flow test. Using a "rule of thumb," they were then to calculate the required fire flow for the property. Material developed during these inspections is kept on file at each station and in our field operations office. Additionally, any fire code violations noted are forwarded to the Fire Safety Division for review. Referrals may also be made to Building, Housing, and Zoning Inspection Departments on referral forms. We now have over 300 targets identified and preplanned through the Fire Management Area system. A computer program for recall is in operation.

Target hazards are videotaped and shown to all fire fighters and officers on a scheduled basis. Aerial maps are also used in pre-fire planning training.

We continue to develop data and reinspect each year. In addition, fire losses are being tabulated by sub-management areas. This allows the individual company, and by shift, to understand more about their own fire problem, as well as the comparison to other areas of the city. This system has already produced several desired effects.[1]

Intensive Inspection Program _____

Long recognized as a leader in fire prevention, the Cincinnati, Ohio, Fire Division has maintained an outstanding inspection program for many years. The origin of this program is described in Chapter 7; however, procedures used are worthy of study and emulation by other cities. The Fire Division inspects all buildings, including homes in the city. The system is described by the Cincinnati Fire Division as follows:

BLOCK INSPECTION

Fire Companies, weather permitting, take their apparatus to predesignated locations and while one member stays with the apparatus and remains "IN SERVICE" by two-way radio, the rest of the company members inspect in the vicinity under the supervision of the Officer In Charge of the company. Companies assigned to this program rotate according to the District Marshal's schedule. This is done so that all companies participate, but in such a fashion that adjoining companies are not out inspecting at the same time. This program called the Unit Block Inspection Program has enabled the department to increase the total number of inspections and allow a heavy concentration of effort in residential areas.

With this method the company is never out of service for fires or other emergency runs. This type of inspection also gives our citizens a good opportunity to view the apparatus and ask questions relative to their Fire Department. They may also receive fire safety instructions and literature from the member who stays with the apparatus. The children get a first-hand view of "the big red fire engines." Experience has shown that in addition to performing a valuable service, the program creates good public relations in the neighborhood being visited.

TACTICAL INSPECTION UNIT

When the frequency of fires has increased critically in certain areas, a task group of fire inspectors in a concentrated inspection program is deployed to reduce hazards. These fire inspectors are detailed from various companies throughout the City under the command of a District Supervisor and concentrate on fire hazards within the critical area until they are rectified.[2]

Cincinnati's fire prevention code enforcement efforts are directed toward serious violations where legal action is indicated. The Fire Division also is responsible for preventing environmental insult and has an environmental crimes unit to carry out these duties. The Fire Prevention Bureau is now the Environmental and Safety Services Bureau.[3]

Canadian fire fighters are also involved in fire prevention activities. As an example, 29 departments in the greater Montreal area are using this technique. Montreal's program is described as follows:

> In addition, firefighters continue to carry out their prevention visits to various districts of the city. These visits provide the opportunity to exchange information with the public and detect any situations which may contribute to increasing the risk of fire.
>
> These various types of public contact enable firefighters to exercise their leadership among the public in the field of promoting self-protection against fire (smoke detectors, fire extinguishers, etc.).
>
> In this regard, since 1987 each shift of firefighters is required to con-

tribute twenty-four and one-half hours of fire prevention work per week, over and above regular public fire prevention campaigns conducted by about forty inspectors, many of whom are women.[4]

Personnel Trends

A number of cities are using non–fire service personnel for fire prevention duties. Dallas, Texas, for example, recruits people with at least two years of college directly into the fire prevention bureau. Their career ladder is limited to the fire prevention bureau supervisory positions, however. These individuals receive 530 hours of specialized training in subjects related to fire prevention.

Many fire departments are realizing the tremendous potential for women in fire prevention work. In some cases women are being assigned to fire prevention duties after completion of basic fire department recruit training. These individuals are members of the uniformed fire service. In other communities women have been employed as civilian fire inspectors or fire prevention education specialists.

The use of women for fire prevention duties is not new. The 1923 Convention of the Oklahoma State Firemen's Association in Oklahoma City was addressed by Olga Juniger of the Texas State Fire Marshal's Office on the subject "Fire Prevention Work in Women's Clubs."[5] She was assigned full-time to fire prevention duties in her home state.

State Responsibilities

State agencies may exercise control over local fire code enforcement activities. As an example, Wisconsin law provides for local enforcement of the state's fire codes; however, the state provides funding for the local jurisdictions to perform the necessary inspections. The state Department of Industry, Labor, and Human Relations employs several inspectors who verify that the local municipalities are actually conducting required inspections. Funding is terminated if inaction is found.[6]

The Relation of Fire Prevention to Other Municipal Agencies

Certainly, the fire department is the major municipal agency with an interest in fire prevention, and it is the one responsible for the overall control and enforcement of fire safety measures. As discussed briefly in Chapter 10, a number of other municipal agencies have a corollary interest in fire prevention as part of their major functions. To complete the picture of the place of fire prevention in municipal government, it is worthwhile to discuss the ways in which the entire municipal government is involved.

The municipal water department has a clear-cut tie-in to fire protection. Any improvements in the municipal water supply and distribution system will result in improvement in the overall fire protection in the community. For this reason the fire chief or fire marshal and the director of the water department should work closely together. Plans for expansion of the water system should always include fire protection considerations. As an example, a six-inch main might be adequate for domestic service in a given section of town, while an eight-inch or ten-inch main may be far more desirable from a fire protection standpoint. If the fire department cooperates with the water department, fire protection needs can be taken into consideration at the time of original development. Plans should also be coordinated if the water system is privately owned.

Another area of cooperation between the fire department and the water department is in testing of hydrants to determine appropriate pressure and volume as compared with those actually available. Coupled with this activity may be the maintenance of fire hydrants, including painting and color coding.

Coordination with the police department is essential for effective fire prevention. In some cities there has been real or imagined friction between police and fire personnel, which may stem from rivalry or from misunderstanding of duties and responsibilities. Regardless of the reason, certainly, the public loses when misunderstandings mitigate services rendered. On the other hand, the public stands to benefit greatly through close cooperation between police and fire personnel in the interest of fire prevention.

In the course of their normal duties, police officers may observe conditions that should be reported to the fire prevention bureau for appropriate action. Overcrowding in places of assembly, for example, is more likely to come to the attention of the police than to the fire service. The police department can also be of great assistance in enforcing regulations covering the transportation of hazardous materials. Police officers are in a better position than the fire department to check on vehicles transporting dangerous materials to ensure the use of required markings, the use of designated parking facilities, and the observance of other safeguards. On the fireground, police and fire suppression forces traditionally work well together. The same kind of cooperation in other areas can advance the cause of fire prevention in the community.

The ultimate coordination of police and fire services is the combination police–fire department mentioned earlier. Regardless of the pros and cons of the overall plan, it may safely be said that it is a mistake to expect personnel who are looked on primarily as police officers to carry out fire prevention responsibilities. The fire service has depended on a low-key public image to help in implementing fire prevention programs. The psychological impact of a person uniformed and armed as a police officer may create the opposite effect. One of the major advantages of the combined police–fire program is said to be the reduction of idle time. It can be argued that a fire department that is carrying out a wide variety of fire prevention and emergency medical programs has little time for police duties.

The local health department will probably have some responsibility for the inspection and approval of facilities from a sanitation standpoint. In the field of restaurant inspection, there is room for a great deal of cooperation between health and fire department personnel. The same may be true in inspection of day-care centers, boarding houses, hotels, and numerous other facilities where health department inspection and approval are required.

Coordination with the building department is in review of plans and specifications and cooperation in inspection for occupancy of structures within the community. Such coordination is extremely important. It should begin with coordinated reviews of preliminary plans and follow through to include cooperative inspections of occupancies. There are many examples of close cooperation between the two agencies. Traditionally, there have been misunderstandings between the two departments (building and fire) in some communities. Such misunderstandings and conflicts can result in compromising public safety—with dire results.

In some communities the building department has been placed under the control of the fire chief. Charlottesville, Virginia, is an example of a city in which the building official's functions were transferred to the fire department. A battalion chief heads the Building and Life Safety Division of the fire department. The process of consolidation was described as follows:

1. The city initiated a review of "downsizing" the organization to provide a more lean and efficient city operation. The intent was to consolidate where common functions could be accomplished.
2. Reviewing the organization, we were able to identify the area of building inspections as a common area that could be enhanced on the life safety aspects as well as provide a continuous and consistent means of reviewing a building's progress through the various stages of construction.
3. There had to be a change in the City Code redesignating the Building Official and placing that position under the Fire Department's umbrella of authority.

4. Ultimate goals of the consolidation.
   - To place all aspects of building construction under the responsibility of one organization.
   - To eliminate inconsistencies of the review process.
   - To enhance the importance of life safety throughout the entire building process.
   - To identify building codes that may influence fire behavior in a negative way and make appropriate code changes to correct any deficiencies.
   - To involve the fire department's personnel in pre-fire planning from day 1.[7]

Winter Haven, Florida, Champaign, Illinois, Kirkland, Washington, and a number of other cities have placed the building department under the fire department. Few have abandoned the system once implemented. The building official was given responsibility for administration of fire services in the consolidation of services in Bismarck, North Dakota.

Fire department responsibility for building code enforcement was addressed in *Municipal Fire Administration,* a publication of the International City Management Association. This text states, "As a matter of strict logic, since most features of building regulation are for fire safety, the building code and fire prevention code could both be made a fire department responsibility. This is the case in German cities where one of the chief officers of the fire department is the city building official."[8]

Building codes, however, contain a number of requirements not directly related to fire safety. A fire chief who assumes responsibility for enforcing the building code must have adequate knowledge to carry out the duties.

Some communities, counties, and states have found it desirable to place all responsibility for enforcement of the fire-related sections of the building code in the hands of the fire department or fire marshal's office, with the building department retaining full responsibility for enforcement of all other sections. This arrangement seems to have worked well. An obligation is imposed on the fire department or fire marshal's office to provide adequate service with competent employees.

In some cities, the fire prevention code is enforced by the building department. All responsibilities relating to building inspection are thus concentrated in one agency.

There should be coordination with the housing department, whether it be a branch of the health agency or a branch of the building division. Housing codes cover many features of fire prevention; so coordination is mutually beneficial. The desirability of joint inspection may be reviewed on a local basis; however, in single-family occupancies, it would not be legal to conduct joint inspections without permission. The fire department may inspect such facilities on a nonobligatory, voluntary basis, while the housing section inspections are obligatory. It would not be in the best interest of either agency to have fire personnel enter on the strength of the housing inspector's right of entry. Both programs might suffer under such an arrangement.

Planning and zoning functions have considerable bearing on fire safety in any community. There should be no hesitation in coordinating fire prevention activities with those of planning and zoning to ensure recognition of necessary storage locations for hazardous materials and to ensure adequate clearances between buildings in the interest of preventing conflagrations. In some communities, the fire prevention chief or fire marshal actually sits as an ex officio member of the planning and zoning board.

The city purchasing bureau, which is generally responsible for the purchase of all supplies for the municipal government, offers an avenue through which the fire prevention bureau can work to have certain fire safety requirements established in city purchasing policies. For example, curtains and draperies and other materials used in public buildings can be specified to meet certain flame-retardant requirements. The purchasing agencies are usually happy to comply with such requests.

The sanitation department has a concern in fire prevention. Disposal of trash and other combustible waste is, of course, a part of the fire problem in any city. The sanitation department can minimize hazards through uniform pickup procedures. The fire prevention bureau and the sanitation department should coordinate their efforts to ensure fire safety during extreme conditions. An example is a special trash collection during times of civil unrest to reduce the possibility of loaded trash cans being set on fire by disorderly persons.

Public utilities may be provided by private or public agencies. Regardless of ownership, fire prevention personnel should deal with these agencies just as they would deal with a municipal department. Areas of coordination include fire code enforcement, cooperation in responding to alarms, and other functions in which mutual interest is indicated. Cooperation with electric power, telephone, cable television, and gas companies is especially important.

The public school system is a government agency with great potential for cooperation in fire prevention endeavors. Cooperation may extend to arranging for fire department staging of fire safety demonstrations, classroom instructions, and cooperation with fire drills and inspections. This type of work has been most rewarding, but it can only be carried out successfully if there is complete understanding between fire prevention and education personnel.

The city library, although not thought of as an agency having a great deal of interest in fire prevention, can be of value by providing space for displays and exhibits and by making available a wide variety of publications in the fire field. Books are available that may instill an interest in fire safety in the reader. For those fire departments that may not be able to pay for an extensive library of their own (including handbooks and reference books on fire prevention matters), the city library may be able to obtain additional books to serve as a valuable supplementary reference source.

The judicial system has a bearing on fire safety. In most states the judicial system is primarily a state operation, although many states do have municipal courts, which are responsible for hearing cases involving housing, building, zoning, and fire code regulations. Thus, the fire prevention inspector may engage in enforcement activities involving cooperation with the courts.

Public welfare programs, which are generally carried out at the state level, offer some opportunities for coordination in fire prevention. A high percentage of public expenditures for fire protection, especially as related to responses to alarms of fire and emergency medical calls, goes to areas with low-income residents. This indicates a need for education in fire safety matters among welfare recipients. Very few jurisdictions have attempted to tie in fire prevention with welfare programs; however, avenues for development should be explored.

Cooperation on the fireground reflects on fire prevention and protection in several different ways. Fires in the community can have a tremendous impact on fire prevention efforts because they draw public attention to the subject of fire and demonstrate the need for fire prevention in a way that should not be underestimated. It is easy for people working in the fire safety field on a full-time basis to get the idea that everyone else thinks about fire prevention most of the time. This, of course, is not true, and many people give no thought to the subject until fire breaks out in their community.

Full cooperation between police and fire departments at the scene of a fire can go a long way toward smoothing out rough edges in public reactions. A motorist stopped by a hose line across a highway is usually irritated at the fire department, especially if the delay is lengthy. Under these conditions, the police department can probably do a better job of handling the public. By treating people in and around the fire scene courteously, they can preserve good public relations and allow the fire team to concentrate on their emergency operations.

Likewise, cooperation between the fire department and the water department is essential during major fires and can benefit both agencies. Deficiencies in water supply that may affect automatic sprinkler protection may show up during the suppression of a major fire. A fire emergency may also show the need for up-to-date information on water supplies, maps, and methods of designating hydrant locations and sizes, water main sizes, etc., as well as the need for augmenting water supplies during major fires and for providing additional hydrants in a given location. All of these factors can be used to press for public support for necessary water supply improvements. This will benefit the community in another way, too. Water supply facilities are recognized as a major factor in the rating of a municipality under insurance rating schedules.

The municipal streets or public works department can lend support to the fire department at the fireground. This support may include the provision of barricades at the scene of a major fire to assist in crowd control.

The streets department may also be able to provide compressors and jackhammers to the fire department for use during major emergencies. In some communities, street flushers are used as auxiliary water carriers by the fire department. Again, such an arrangement in an emergency situation may save the fire department a considerable financial investment.

As important as the fire service and fire prevention functions are to the community, other municipal departments are equally concerned with the safety and welfare of the public.

Summary _____

By tradition, fire prevention and fire protection in the United States have been primarily local government responsibilities. In the charters of most cities, public fire protection, which can easily be interpreted to include fire prevention, is named as a responsibility of the municipality. Court decisions have clearly indicated that the legal responsibility of a municipality to protect its citizens encompasses fire protection.

With thousands of incorporated cities and municipalities in the United States and Canada, the public fire services vary widely, from volunteer and call groups in the smaller communities; combination volunteer and career forces; to large, fully career fire departments with specialized divisions for fire prevention and training. In the traditional form of the municipal fire department, the fire chief reports directly to the mayor or other top administrator. In some municipalities, the fire department is placed under a director of public safety, who is also responsible for the police department. Another variation is the combination police–fire department.

Fire prevention refers primarily to measures directed toward avoiding the inception of fire, among which are fire prevention education, fire safety inspection, fire code enforcement, investigation of fire to determine causes, and investigation of suspicious fires. In fire departments serving communities with populations of 25,000 or more, a fire prevention bureau generally coordinates all fire prevention activities. The bureau may range from a one-person operation to a relatively large operation with full-time personnel assigned to specialized functions, such as educational programs and technical inspections. Fire prevention personnel should be suited to dealing with the public, and they should also have suitable quarters, easily accessible to the public, with adequate communication facilities to carry on their work.

Conflicts may arise between fire suppression personnel and the fire prevention bureau for several reasons: lack of understanding on the part of fire suppression personnel of the nature and importance of fire prevention work; differences in hours; social status associated with automobiles assigned to fire prevention personnel; and a general feeling that the fire prevention people have it easy, don't take the physical punishment that fire fighters do. Some departments are requiring all fire service personnel to spend some time in fire prevention work; others make fire prevention training a part of their basic training; others make fire prevention assignments a requisite of the rank of lieutenant or captain.

Other innovations to advance the cause of fire prevention are fire prevention advisory committees, made up of members from all divisions of the fire department—and sometimes also of people from outside the fire department—to consider and decide on programs and policies; and the mobile fire safety team, which patrols the community in a special vehicle equipped with demonstration fire fighting equipment and educational materials to help in promoting fire prevention awareness among the citizens.

Fire prevention is a concern of other municipal agencies, too, and the fire prevention bureau should cultivate opportunities to work with them in every possible way to get the message of fire prevention across to the whole community.

Notes

1. Personal communication to the author, 1994.
2. Cincinnati Fire Department, "The Cincinnati Fire Department System of Building Inspection" (Cincinnati, 1976).
3. Personal communication to the author, 1994.
4. *Changing for the Future* (Montreal, Quebec: Service de la prevention des incendies de Montreal, 1989), p. 12.

5. Oklahoma State Fire Fighters Association, *The First 40 Years* (Oklahoma City), p. 89.
6. *Wisconsin Administrative Code* (Madison, Wisc.: Department of Industry, Labor and Human Relations, 1993), pp. 4, 5.
7. Charlottesville Fire Department, Consolidation Information (Charlottesville,Va.), 1991.
8. International City Management Association, *Municipal Fire Administration* (Washington, D.C., 1967), p. 232.

CHAPTER 4

Preparing Fire Service Personnel for Fire Prevention Duties

Many members of the general public assume that fire department personnel without exception are motivated in fire prevention work. Unfortunately, probably only a relatively small percentage of fire department personnel, both career and volunteer, are in fact interested in or knowledgeable in fire prevention.

The primary motivation behind the organization of most fire departments has been that of suppression. Many people believe that the fire department's obligations have been met if the department responds to and brings under control all fires to which it is called, and fire prevention is not generally thought of as being a part of the basic responsibility of many fire departments.

Lack of Emphasis on Fire Prevention _____

Training programs in the fire service have likewise had limited emphasis of fire prevention. They have for the most part concentrated on fire suppression subjects. The average individual joining either a career, volunteer, or call fire department has received little information relating to fire prevention in the basic training program. Members of many explorer and other scout groups receive more training in fire prevention than does the person joining the average fire department. Fortunately, the trend in the fire service has been to include fire prevention training as part of the basic program. Several programs directed at the volunteer fire service also include fire prevention training.

Major fire conferences designed primarily for fire service audiences have not been noted for their strong emphasis on fire prevention. Talks on fire prevention have not been found on fire service conference programs to any great extent. When included, they are often programmed during a night session or at some other time when attendance is optional.

National periodicals in the fire service field have likewise given a lesser amount of coverage to fire prevention than to other phases of fire service activities. A change in emphasis is beginning to appear, and periodicals are paying more attention to fire prevention.

There are probably other reasons than those already mentioned why fire prevention has not received its just deserts in the overall fire service picture. Motivation is certainly more limited because of the less glamorous nature of the tasks performed. The excitement and activity of the fire scene are far more challenging and rewarding to many people than is the routine of a building fire prevention inspection or staging a demonstration before a group of six-year-olds.

The fruits of labor are also much more readily recognizable in fire suppression efforts than in fire prevention work. The individual engaged in fire suppression activities is rewarded with direct evidence of something accomplished. The control and extinguishment of a major fire are highly visible, whereas the rewards of fire prevention work are usually intangible and not so easily measured.

An individual assigned to fire suppression duties seldom comes in contact with the public in adverse situations. The fire preventionist is often in an adverse relationship with those dealt with, especially in code enforcement. Citizens do not usually care for advice to extinguish cigarettes, provide automatic sprinkler systems, remove obstructions from exits, and complete other fire safety requirements. They are, in fact, often alienated by such efforts. This is probably another reason for the reluctance by some fire service personnel to accept fire prevention as a major fire department activity.

For years the insurance industry has made contributions in the field of fire safety education. The fire service, however, has the primary responsibility for the prevention of fires. Some fire service personnel believe that the insurance industry is the major benefactor from any efforts they might expend in the fire prevention field. Such a thought is probably based on the fallacious assumption that insurance companies directly profit from a reduction in fire losses, whereas, in fact, insurance rates are adjusted according to actual losses.

Fire Prevention Training for All Fire Service Personnel

In preparing fire service personnel for fire prevention duties, it is important to consider the needs of the individual fire fighter. We should consider the motivation factors.

One important incentive is that fire safety work promotes the professionalization of the fire service because of the higher level of knowledge and skills required to do the job. There is no question that certain skills required for fire prevention assignments are beyond those required for basic fire suppression duties. Technical training and proficiency are needed to understand the hazards of industrial processes and the uses of dangerous chemicals and to evaluate construction methods. Fire prevention training opens up new fields of specialization for fire service personnel and widens their range of activities in the service.

The fire fighter also gains knowledge that is helpful in fire suppression duties. In conducting fire inspections and prefire planning surveys, the fire fighter has an opportunity to learn about conditions in the buildings in the community before a fire occurs. This information can be of great help when entering a building during a fire emergency.

A vigorous fire safety program that places fire service personnel in frequent contact with the people of the community can do much to enhance the image of the fire department. It may help to erase some preconceived notions on the part of the public of the work of the fire service. Too often, the picture of the fire fighter in the public mind is that of an individual leaning back in a chair in front of the fire station, "sitting and waiting for a fire." If the public can be brought around to thinking of the fire fighter as an active public service employee who is on the street preventing fires when not engaged in emergency medical or fire fighting activities, certainly, the overall good is served.

Another result of public contacts by fire service personnel in fire prevention duties, especially if they are favorable in nature, is that, as citizens become more aware of the hazards of fire and gain confidence in the fire service, they could well be encouraged to call the fire department without delay in the event of an incipient fire, no matter how small, instead of trying to cope with it themselves and risking spread. Many communities have found that after a stepped-up fire safety program was put into operation, the number of fire calls was reduced, presumably because fire prevention advice was being followed. There is less likelihood that this would have a major effect on the loss per fire, but any reduction in the number of fires means an overall gain to the community.

Because of the community involvement aspects of fire prevention work and the opportunities they provide to build good public relations, public respect for the fire service is bound to increase. This can have a favorable effect on the ability of the fire department to obtain funding for apparatus, personnel, and new stations. The citizen who is favorably impressed with the fire service is ready to support efforts to increase the department's personnel, improve facilities, and otherwise upgrade the services rendered. Provision of emergency medical service also has this effect.

Recruit Training Programs

An example of an excellent training program in the field of fire prevention is that of the Dallas, Texas, Fire Department. As part of the recruit training school, each new member is required to undergo an additional 56 hours of training in fire prevention and related subjects.

The fire prevention subjects are designed to give the recruit an appreciation of the worthwhile purposes of fire prevention duties. Subjects covered during the fire prevention phase of the training program include objectives of fire department inspections, fire hazards and causes, inspection techniques and procedures, prefire planning and inspection, fire cause determination, and fire protection systems.

The New York City Fire Department likewise has a program for indoctrinating new personnel in fire prevention methods. It includes fire prevention training at the fire academy and in the station as part of its overall program.

Louisville, Kentucky, includes 80 hours of class on fire prevention and public education in its 720-hour recruit training program. Subjects include methods of conducting home and nonresidential building inspections, technical fire codes, home fire safety lessons, and public fire education. As a final examination, recruits are required to conduct fire safety classes for high-risk groups in the city.[1]

Many cities assign fire suppression companies to conduct code enforcement inspections. Training must be provided to participating fire company personnel. Some states require instructors as well as inspectors to be certified. In Salina, Kansas, for example, fire prevention officers are trained and certified by the state as instructors. They, in turn, instruct all company personnel in basic fire prevention inspection and cause and origin determination. All fire fighters and company officers are certified by the state as inspectors.

Prefire Planning

There is a very close relationship between prefire planning and structural control inspections. Some fire departments have attempted to combine these visits; however, this is not advisable because of the differing purposes of the activities.

Prefire planning surveys are primarily carried out for the purpose of familiarizing fire department personnel with building locations and floor plans as well as sources of potential danger or other unusual features. The surveys also provide fire suppression personnel with knowledge of the locations of sprinkler control valves, standpipe connections, and other information relating to the fixed fire protection equipment in the building. The location of hazardous commodities that may endanger fire fighting personnel is also noted. Information obtained in a prefire planning survey must be available for personnel responding to emergencies. This information has proved to be most helpful to fire suppression personnel on the scene of a fire. No doubt many fire fighters have avoided death and injury because of prefire planning programs.

Prefire planning surveys and fire prevention inspections are devised for two different purposes, and combining the two into one inspection may place the property owner in a rather difficult position. The owner is interested in providing fire suppression personnel with information helpful in fighting a fire but is probably not as interested in having fire personnel enter for the purpose of bringing about corrective action. If the same fire fighters are assigned to carry out both responsibilities at the same time, there is a possibility that the two undertakings will conflict with each other. If both fire safety inspection and prefire planning visits are to be carried out at the same premises, there must be close coordination in scheduling of inspections and visits. The program has excellent public relations value.

In some jurisdictions the word *planning* has been deleted from the title of this program because of potential legal liability in the event that fire fighting efforts are unsuccessful despite the presence of a "plan." In other locations the term *prefire survey* is used; however, this may cause problems because it presumes that there will be a fire. Regardless of the title used, training of personnel performing these duties is imperative.

Fire Prevention Training Assignments _____

Another approach is to assign personnel to fire prevention functions at some time during their career in the fire department. In several departments, for example, personnel promoted to the rank of lieutenant must, during their tenure in this rank, serve in the fire prevention bureau for a specified period of time. This procedure has been found to be quite helpful. An individual moving up through the rank structure gains an appreciation of fire prevention. Other departments make it a practice to assign individuals to fire prevention while they hold the rank of captain. Both of these practices do a great deal to preclude the possibility of an individual reaching the position of chief officer without having served in fire prevention.

Greater emphasis is being given to the use of all active fire service personnel for fire prevention duties. A time may come when fire prevention will be a recognized duty assignment for anyone entering the fire service, as is the case with forest fire fighters. Such an approach will go a long way toward reducing fire losses.

Georgia's Fire Commissioner appointed a 60-person commission to study ways to reduce injuries, deaths, and property damage caused by fire. Members representing both public and private sectors held hearings in 12 Georgia cities to obtain input. The Commission was divided into four committees: Training and Operations, Codes, Investigations, and Public Education. Their final report was issued late in 1993.

A major recommendation of the Codes Committee was that provisions be made to permit building officials to be trained at Georgia's Public Safety Training Center, which is the state's fire training center. The report points out the advantage of having joint training of building and fire officials.

The Public Education Committee recommended that public fire safety education be made a part of basic minimum standards training for both career and volunteer fire fighters. They also recommended that newly elected local public officials, as well as newly appointed fire chiefs, receive indoctrination in public fire safety education.[2]

South Carolina's Fireman's Association supports a Fire Safety Cadet summer camp near Aiken each summer. The state's 4-H Club, through Clemson University, and Aiken's Department of Public Safety are also participants in operating this week-long camp for 12- and 13-year-olds from across the state. Classes and activities emphasize fire safety and may also provide campers with an incentive to eventually become members of the fire service.

Training in Fire Prevention Inspection

In addition to the indoctrination and training of general fire service personnel in fire safety duties, there is a need for training in inspection procedures for full-time personnel in the fire prevention bureau.

As an example of early training in fire prevention duties, the following instructions were given to fire inspectors in the Philadelphia Fire Department in 1913:

Major fundamental principles in the training of fire inspectors:

1. The work of fire prevention may impress you as a fad or a joke at the outstart, but your matured view will be that it is both interesting and serious and in comparison with fire extinguishing it stands as a Bureau of Police to the Detective Bureau.

2. Your sound judgement of corrections to remove fire dangers still has the backing of your superior officers against all protests, and to secure the desired result, you are to permit no hindrance in your work through fear or favor, friends or enemies.

3. Be courteous, explain in detail your mission and do not quarrel with the occupants of any building. Remember there is a distinct advantage over the man who has lost his temper to the man in an argument with self-control. You are representing the dignity of a department of your city and must in no case initiate a quarrel; if you cannot, report the facts to the officers in charge of the fire prevention work and have them open this avenue for you.

4. Make friends with the engineer, janitor, superintendent or any other person in charge of the building. These men have considerable information of the property condition and influence in having corrected promptly conditions which you desire improved.

5. This new responsibility calls for intelligence, sound judgement, broad views and being amenable to reason. Narrow-mindedness and bombast have absolutely no place in this work and every new and practical idea bearing on the end ultimately to be achieved should be promptly grasped and utilized.

6. Gently and firmly decline the friendly cigar or other small article usually offered in good faith, but occasionally used by occupant or owner of property to put the inspector in a bad light. In particular, guard against the least suspicion or suggestion of graft in this form and by adhering to strictly business dealings you are in a stronger position should your work be questioned or reviewed.[3]

These principles are as applicable today as they were in 1913. They should be considered and adhered to by everyone assigned to fire inspection duties.

The more an inspector knows, the more valuable are the inspections. A major problem in effective fire prevention inspection is that of failing to see what is being looked upon. This failure may be the result of lack of training in fire safety inspection, or it may simply be the individual's inability to register problems seen during the inspection. An inspector may report that a complete inspection has been made by going around and checking on fire extinguishers and on enforcement of smoking regulations, when in reality an effective inspection has not been made.

Educational needs include training in inspection procedures, familiarity with fire codes, training in plan reviews, an appreciation of building code fire protection features, an understanding of health codes as related to fire safety, as well as an understanding of court decisions affecting the field. The fire inspector should also have a general knowledge of procedures to determine the causes of fires.

In smaller fire departments, where it may not be practical to establish formal programs, training procedures often consist entirely of on-the-job training. With this method it is essential that the trainee be assigned to work with an inspector who is fully qualified to perform inspection duties. Such training might take place in another city with a larger bureau, or at a state fire academy. The National Fire Academy also offers extensive training in this field.

Many larger departments have formal training programs for fire safety inspectors. The case study method is often employed in these programs. Case histories of actual field problems in fire prevention and fire code enforcement are studied and analyzed as part of the training program.

The fire safety inspector should have available reading material pertaining to changes in the field. All publications, including books and magazines relating to new processes, new fire protection equipment, etc., should be routed through fire prevention inspectors. Weekly conferences and training sessions are also helpful.

Although technical fire inspection duties require more extensive training, a great deal of effort is not always necessary to prepare fire service personnel for basic fire prevention duties. Some programs, including home inspection and basic fire prevention education, can be made effective with relatively simple training methods.

National Professional Qualifications System _____

The Joint Council of National Fire Service Organizations, a group composed of the major national fire service organizations, agreed in 1972 to establish standards for professional competency in the fire service. The Joint Council no longer exists. Committees were formed to develop and prepare recommended minimum standards of professional competence for four groups: fire fighters, fire inspectors and investigators, fire service instructors, and fire service officers. The composition of these committees was based on Joint Council membership.

The National Professional Qualification Board was established by the Joint Council of Fire Service Organizations to oversee and validate the standards developed by the four committees. The board was also responsible for establishing and maintaining a nationally coordinated continuing professional development program for the fire service. In 1990 responsibility for oversight of the Professional Qualifications project was assumed by the National Fire Protection Association. The committees are established and operated under the National Fire Protection Association's standards-making procedures, and proposed standards and revisions are submitted to the Association for final adoption.

The standards may be used by states and communities as part of their requirements. The system is designed to encompass all duties of the uniformed fire service member as well as certain "civilian" positions.

It was proposed that each state have a certification process. The National Board on Fire Service Professional Qualifications accredits state certification boards and certifies individuals. The International Fire Service Accreditation Congress also accredits state programs. Certification at each level, starting with fire fighter I, would result from the individual's successfully completing certain performance objectives. Time limitations are not imposed, nor are minimum hours of training or years of service. The individual's progress is based on the ability to master requirements and demonstrate proficiency to an examiner. Of course, promotions would be limited by the billet structure of the fire department.

The professional qualifications system does not attempt to assign ranks to levels of qualifications. It envisions that local jurisdictions would make these determinations. A wide range of interests are represented by the National Fire Protection Association committees, which develop the standards.

Although designed primarily for the career fire service, the term *professional* in this context connotes individual abilities rather than employment status. Volunteer and combination career and volunteer departments can participate in the system; in fact, many are already utilizing the standards.

The system envisions that training programs will prepare fire service personnel to acquire skills and knowledge necessary to achieve the performance objectives of the standards. It also envisions a controlled testing procedure for candidates.

The current standard, *Professional Qualifications for Fire Inspector,* National Fire Protection Association (NFPA) No. 1031, was designed to ensure the competency of personnel assigned to these duties.[4] The standard is not designed to be used for personnel other than those in the fire service. It does include "direct entry" individuals, however.

The original standard developed by the Committee on Fire Service Professional Standards Development for Fire Inspector and Investigator Qualifications was known as *Fire Inspector, Fire Investigator and Fire Prevention Education Officer Professional Qualifications,* NFPA 1031.[5] This standard included requirements for progression in all three specialties. An individual was required to have been certified as a Fire Fighter III before progressing to the entry level in the fire prevention series, Fire Inspector I. The system provided for progression to Fire Inspector II and Fire Inspector III. An individual could start in the fire investigator series as a Fire Investigator II and then progress to a Fire Investigator III.

Likewise, the fire prevention education series began with a Fire Prevention Education Officer II and progressed to a Fire Prevention

Figure 4-1   Fire investigators need to be skilled in photography. (El Paso Times.)

Education Officer III. In both the investigator and education officer series, it was necessary for an individual to qualify as a Fire Inspector I before moving into the specialist area.

The updated procedure as approved in 1987 divides the original standard into three standards, NFPA 1031 (*Professional Qualifications for Fire Inspector*); NFPA 1033 (*Professional Qualifications for Fire Investigator*); and NFPA 1035 (*Professional Qualifications for Public Fire and Life Safety Educator*). This arrangement provides qualification requirements for individuals entering directly into the three specialist areas. The growing tendency to employ individuals specifically for assignments in fire investigation and public fire safety education brought about this change.

Full use of the fire inspector, fire investigator, and public fire educator billet structure is found only in a larger fire department. Hundreds of smaller fire departments assign only one person to fire prevention, investigation, and education duties. In these communities, the one individual would need to meet the requirements for all of the duties imposed.

Two of the three duty assignments may be combined in some fire departments; that is, the same person may carry out inspection and investigation duties while another person performs education duties. In a number of communities, inspection and education duties are combined with a separate person or persons performing investigation duties. In either event the individuals so assigned would need to master the requirements for their duties. The standards do not envision that a community will necessarily establish separate positions for each job classification or level of progression. Some communities or states may desire to impose higher standards than those included in the NFPA standards.

Candidates for progression under the standards are required to demonstrate an ability to express themselves clearly orally and in writing. They are also required to demonstrate their ability to interact with the public under scenarios of code enforcement, fire investigation, or fire safety education duties with tact and discretion. The candidate is also required to demonstrate a knowledge of personal safety practices and procedures.

People aspiring to be fire inspectors at level I must demonstrate knowledge and abilities in many specific areas, including flammable and combustible liquids; compressed and liquefied gases; explosives; heating and cooking equipment; principles of electricity; safety to life; code enforcement procedures; fire cause determination; and miscellaneous provisions. The fire inspector II is required to master these subjects in greater complexity and to be able to develop plans relating to fire safety. As previously mentioned, the fire inspector III must also master administration and management functions.

In states where fire investigators are considered to be officers of the law, they are required to meet applicable minimum standards set for law enforcement officers. They must also demonstrate knowledge and abilities in areas including legal, evidence collection, photography, fire scene examination, records, reports and documents, and courtroom procedures.

The public fire and life safety educator must demonstrate knowledge and abilities in general areas and in public information and public education fields. The individual must have teaching abilities in order to fulfill these requirements successfully. He or she must also understand plant and institutional emergency organization and have associated instructional abilities. The public fire educator III also must master administration and management functions, including an ability to develop examinations and evaluate personnel skills.

As a specific example, the fire inspector I must be able to compute the allowable occupancy load of a single-use building or portion thereof, given a detailed description of the building or portion of the building, so that the calculated allowable occupant load is established in accordance with applicable codes and standards. The public fire and life safety educator III must create a program budget, given organizational goals, community needs, and budget guidelines, so that overall program needs are met within budget guidelines.[6]

There are many reasons for the adoption of professional qualification requirements for fire safety personnel. The system helps develop qualified personnel and goes a long way toward assuring the city administration and the public that fire prevention, investigation, and education duties are being carried out by qualified individuals. Several state court decisions provide an impelling reason for ensuring that personnel are fully qualified. The necessity for courtroom qualification of fire investigators as expert witnesses makes the system almost imperative for these people.

A growing number of states are requiring individuals who perform fire safety inspections to be certified under a state program. Some states also certify fire investigators and public fire and life safety educators. In these states the NFPA standards are generally the basis for the certification requirement.

Many jurisdictions use fire suppression personnel for fire inspection duties. In states with certification requirements, this means that all fire fighters must be certified as fire inspectors. Fire fighters are carrying out the majority of fire inspections, and fire prevention officers serve as resource specialists and handle complex fire safety assignments.

Many other fields of endeavor have had professional qualification systems in effect for years. Included are public accountants, insurance underwriters, electricians, barbers, and a variety of categories of employment. The public can have reasonable assurance that a barber is adequately trained but cannot always have that assurance with a fire inspector. Under the professional qualifications system, a fire inspector I in New Hampshire has basically the same abilities as a fire inspector I in Arizona.

Approximately 390 two-year fire science programs may be found in community colleges in the United States. Many of these programs offer courses in fire prevention and fire investigation. Attendance is primarily from public fire departments. However, building inspectors, architects, and engineers may be found in these classes as well. Most fire science programs offer courses in building and fire prevention codes and standards.

Many fire departments are encouraging fire prevention bureau personnel to enroll in fire science programs. Time off to attend classes or special financial stipends may be offered as inducements for enrollment.

The four-year fire protection and safety engineering technology program at Oklahoma State University, Stillwater, has been looked on as a leader in academic fire protection programs. Since the program's inception in 1937, hundreds of graduates have had a substantial impact on fire prevention and fire safety in many countries. Graduates obtain a bachelor of science degree in Fire Protection and Safety Technology.

Over a dozen colleges and universities now offer bachelor's degrees in the fire protection field. The University of Maryland, College Park, is viewed as a leader in offering fire protection as an engineering curriculum.

Summary

The primary motivation behind the organization of most fire departments has been the suppression of fire in the interest of public safety, and little thought has been given to the allied function of fire prevention. Training programs for fire service personnel have concentrated for the most part on fire suppression activities, and there has been a conspicuous lack of emphasis on the prevention of fire, even in the programs of the major fire service conferences and in the national periodicals that serve the fire protection field.

The trend of late has been toward recognition of the rightful place of fire prevention as a twin function of fire protection. Major fire departments are beginning to include fire prevention training as part of their basic training programs, more attention is being paid to the subject of fire prevention by writers and periodicals in the field, and many city fire charters specifically name fire prevention as a public responsibility.

Fire prevention training for all fire service personnel can have beneficial results in many ways. It promotes professionalization of the fire service because more knowledge is required than for basic fire suppression duties, and it opens up new fields of specialization for growth on the job for all fire service personnel. It also helps to generate respect for the purposes of fire prevention.

A vigorous fire prevention program can also do much to enhance the image of the fire department in the community. If the public can be brought around to thinking of fire service personnel as active public service employees who are on the street preventing fires when not engaged in fire fighting or emergency medical activities, certainly, the overall good is served. And the citizen who is favorably impressed with the fire service is ready to support efforts to obtain funding for needed apparatus, equipment, new stations, and other fire department improvements.

Among fire departments that have introduced fire prevention subjects in their recruit training programs are the Dallas, Texas, Louisville, Kentucky, and New York City departments.

Specialized training needs for fire prevention inspectors include a thorough grounding in the local fire code and procedures in conducting an effective inspection; an understanding of other related codes, such as building codes and health codes; familiarity with fire investigation procedures; and recognition of the legal aspects of inspection responsibilities. Many of the larger departments have formal training programs for fire prevention inspectors; smaller departments may employ on-the-job training. One of the earliest training programs for fire inspectors on record is a 1913 program of the Philadelphia Fire Department, which outlined principles that are surprisingly applicable today.

Although the more technical fire inspection duties require more extensive training, basic fire safety education and home inspections can be made effective with relatively simple training methods.

An important development was the establishment of a National Professional Qualifications System for state certification of fire service personnel at several levels of professional competence in the classifications of fire fighter, fire inspector, fire investigator, fire service instructor, and fire service officer. Standards were originally developed by committees appointed by the Joint Council of National Fire Service Organizations with staff support by the National Fire Protection Association. Adoption of these standards is a major step in professionalizing the fire service and improving the nation's fire protection capabilities.

Notes _____

1. Russell E. Sanders, "Chiefs Briefing," *NFPA Journal,* National Fire Protection Association (November/December 1993), p. 14.
2. Commission for a Fire Safe Georgia, *Goals for a Fire Safe Georgia* (Atlanta, Ga.: Office of Safety Fire Commissioner, 1993) pp. 13, 22.
3. Powell Evans, comp., *Official Record of the First American National Fire Prevention Convention* (Philadelphia: Merchant and Evans Co., 1914), p. 150.

4. *Professional Qualifications for Fire Inspector, NFPA 1031* (Quincy, Mass.: National Fire Protection Association, 1993).

5. *Fire Inspector, Fire Investigator and Fire Prevention Education Officer Professional Qualifications, NFPA 1031* (Quincy, Mass.: National Fire Protection Association, 1977).

6. *Professional Qualifications for Fire Inspector, NFPA 1031,* 1993.

CHAPTER 5

Fire Safety Inspection Procedures

Inspection is a key function in the enforcement of fire laws and regulations. Duties of the fire prevention inspector call for knowledge and competency that can be acquired only through proper training and education in code requirements and inspection procedures. Other requisites are the ability to exercise good judgment, keen observation, and skill in dealing with people. As a representative of the fire department, the fire prevention inspector has the opportunity to build good public relations as well as to educate the public as to the need for observing the rules for prevention of fire.

This chapter outlines the basic principles that apply to the actual inspection procedures, whether conducted for control of structures, for control of occupancy, or for a combination of purposes. For more detailed information regarding inspections, excellent sources are publications of the National Fire Protection Association and the International Fire Service Training Association.

Preparation for Inspection

Preparation should include the instilling of a positive attitude on the part of the inspector. The inspector must know why the inspection is being made and what to look for during the inspection. This is an important phase of preparation, one that, unfortunately, is often overlooked in training and in planning for inspection programs. Another requirement is the acquisition of necessary equipment, such as flashlights, cameras, notebooks, and suitable clothing. In some departments the inspector is also required to carry manuals, code books, and other publications that may have a bearing on the work. It certainly is a good idea to have these materials in the car for reference if needed.

Knowledge of Codes

One of the prime requisites for competence in the inspector's duties is a thorough knowledge of the local fire safety code, as well as other relevant municipal or state codes that might apply. The inspector may not necessarily be fully conversant with each point in the fire prevention code but should be familiar enough with the provisions and requirements to make general observations and recommendations during the inspection tour, with subsequent detailed study where necessary. This might be done at the site by reference to code material carried in the car. In some instances, special problems may require more detailed analysis or consultation with supervisors after the inspector has returned to the office.

Time Requirements

Under no conditions should an inspection be undertaken when the inspector is pressed for time and is thus unable to make a thorough inspection. Fire safety inspections made under pressure of time would in most cases better serve the cause of fire prevention had they not been made at all, because an official inspection, however haphazard, may lull the owner or property manager into a false sense of security and subject the inspector and fire department to legal action.

Although several major insurance organizations have experimented with time quotas for conducting fire prevention inspections, this is a risky practice to follow. The complexities of building inspections and the varied degrees of understanding on the part of the individual accompanying the inspector, as well as a number of other factors, combine to make it most difficult to stipulate a fixed amount of time for a fire inspection. An estimate of the average time needed for a given occupancy might be established, but under no conditions should an effort be made to limit the inspection to that figure. Thoroughness cannot be overemphasized. It is necessary to cover adequately all areas of the structure even though entry to certain spaces may be difficult. In many buildings, there may not be a master key for all closed areas, and certain spaces may be assigned to personnel who carry individual keys. This necessitates a search for the individual who has the specific key, which slows down the entire procedure. It is essential to take the time to obtain entry to all spaces.

Inspection Frequency

Inspection frequencies vary considerably from one jurisdiction to another. Availability of personnel is a major factor in determining the amount of time between inspections. Another factor is the perceived degree of hazard in the occupancy. For example, institutions are usually inspected more frequently than storage warehouses. The potential for finding deficiencies may be a factor in scheduling inspections even within a given class of occupancy. Those facilities in which violations are usually found may be scheduled for more frequent visits than those where fire safety problems are seldom found.

North Carolina has mandated that inspections be conducted on a scheduled basis. Subsequent to the Hamlet food processing plant fire, on September 3, 1991, the state's Building Code Council added a mandatory fire safety inspection schedule requirement to the Fire Prevention Code. Local governments are responsible for conducting these inspections of occupancies.

Hazardous, institutional, high rise, assembly and common areas of residential occupancies must be inspected at least annually. One- and two-family dwellings and townhouses are exempted. Industrial and educational occupancies except public schools must be inspected at least once every two years. Public schools were already required to have at least two fire inspections per year, spaced no less than 120 days apart. Business, mercantile, storage occupancies, churches, and synagogues require inspections at least every three years.[1]

Another measure adopted subsequent to the Hamlet fire was the State Department of Labor's adoption of the National Fire Protection Association's (NFPA's) Life Safety Code as part of their regulations. Most places of employment are covered by these provisions.

Inspection Preliminaries

Before leaving the office or fire station, the inspector should look over reports of any past inspections at the facility. These reports will be of material value in conducting an inspection and must be considered in evaluating the facility. A property owner is certainly justified in being alienated by a fire inspector who fails to recognize past efforts to correct deficiencies. Most property owners are proud of efforts they have made to correct deficiencies, and their efforts should be recognized. By reviewing past reports, the inspector can also get a better idea as to the construction and general layout of the building, and the time taken to review the file will be rewarded by making possible a more expedient inspection.

If the previous inspection was made by another inspector, it may be necessary to confer with that individual to ensure a clear understanding of all items listed in the inspection report. If the inspector going out on the job finds some previously reported items confusing, the property owner is most likely also having difficulty in understanding what is meant. Furthermore, the last inspector may have missed some items.

Manner of Dress

There is some interest in plainclothes dress for fire safety inspectors. On entering the premises, a person so dressed is not as apt to create a feeling of resentment on the part of the owner or operator as is someone in uniform and is not likely to be noticed by customers, a pleasing factor to the proprietor. Some form of identification is of course necessary. Appropriate identification cards or a blazer with an emblem or badge may be used. Many departments feel that a uniform is more effective in gaining compliance, however.

Taking Notes

The inspector should be prepared to carry a notebook, clipboard, recorder, or computer to make notes of all findings during the inspection. In any case, the notes taken should be orderly and easily understandable after a full day of inspecting many types of facilities. Specific locations of items needing correction should be designated in the notes, rather than general statements that might prove difficult to understand. There should never be an indication to the inspector's escort that notes are being kept in a slipshod manner. If an inspector uses the back of an envelope or a matchbook cover for notes, it gives the guide the impression that a thorough, accurate inspection is not being made.

Identification and Permission to Inspect _____ ____

The approach to and contact with the owner or operator of ι property are important factors in establishing a favorable climate in which to conduct the inspection. On entering the structure the inspector should immediately present identification to personnel in the building. If at all possible, the manager, superintendent, or other person in authority should be notified. Such identification should be made to ensure compliance with the guidelines of the U.S. Supreme Court decisions in cases pertaining to this point (*See v. City of Seattle* and *Camera v. City and County of San Francisco*). The decisions handed down in these cases make it imperative that the fire inspector present adequate identification to the proper personnel and request permission to make the inspection.

The cases referenced substantiate the property owner's right to refuse the inspector entrance to nonpublic areas except in situations involving imminent danger. The inspector must obtain a warrant to enter nonpublic areas when refused permission to enter.

On occasion a delay will be encountered because those in authority at the property may not be available at the moment the inspector

Figure 5-1 Fire inspections include checking automatic sprinkler systems. (Flagler County, Florida, Dept. of Emergency Management.)

arrives. The inspector should await the arrival of the person in authority and then present identification and request permission to conduct the fire prevention inspection.

Under no circumstances should the fire inspector proceed to make an inspection without proper identification and permission. It might be tempting to go ahead and make an inspection with the agreement of a janitor or other person not responsible for the facility. This can prove embarrassing, because it is generally looked on by management as an intrusion and may even result in management's taking punitive action against an unauthorized person for permitting the entry of the inspector.

When permission is granted to make the inspection, the customary procedure is for the inspector to request that a representative of management go along on the inspection tour. In most cases management readily agrees to this procedure.

The Inspection Tour

For many years it has been customary to start an inspection at the top of the structure. This should include a walk out on the roof to check any roof structures with respect to fire safety. The inspector should check the elevator machinery room and should note whether the roof is used for occupancy. If it is, a check should be made to ascertain that proper egress exists.

The inspector and the escort (or escorts) should observe the condition and construction of the roof as well as make note of any fire protection problems that may exist at that level. This is also a good opportunity to view adjacent structures to evaluate any exposure hazards to the structure being inspected.

The inspection party should then move to the top floor of the building and make a thorough inspection of each section of the floor. Progress through the building should be systematic, and the same principles should be applied down through the entire structure, including the basement levels.

The inspector should avoid any diversionary tactics that might be used to keep him or her out of certain areas. There may be an effort to steer clear of hazardous areas or those in which conditions may not be found up to standard. Attempts to detour the inspector may not necessarily be made to avoid exposing hazardous conditions but may be based on a desire to keep secret certain proprietary processes, machinery, or research work. Industry often has a great deal at stake in the development of new processes or products. An unthinking fire inspector can cause a great deal of hardship to an industry by passing on such proprietary matters. Escorts may need to obtain special permission for the fire inspector to go into certain spaces in which these processes are being carried out. This may mean a delay; however, it is important that the inspector wait.

In making the tour through the building, the inspector should make a point to think about everything noted. One of the major weaknesses in fire inspections is that many inspectors look at hazards, unusual conditions, or exit deficiencies but somehow what they see does not register in their minds as a problem. This is not necessarily a result of inadequate training in inspection procedures; understanding the significance of what is observed is a faculty that takes practice to develop.

The building must be considered from the standpoint of the possible development of a fire within it and from the standpoint of potential spread of fire. Careful consideration must also be given to anticipated occupant reaction in the event of a fire.

Conversation with escorts can cause the inspector to overlook important items. If the person going around with the inspector is carrying on an extended conversation about sports activities or television programs, for example, the inspector can easily walk by a hazard without taking notice.

Although it is impossible to determine every factor that must be considered in examining a building, the salient points for inspection may be classified as follows: exposure hazards, potential fire hazards outside and inside of the building, potential fire causes, potential avenues for spread of fire, water supply, life safety features including means of egress, and extinguishing systems.

Possible Exposures

The inspector should take full note of all appurtenances to the building, as well as exposures and hazardous locations nearby. Time spent in looking around outside to size up the proximity of other structures or combustibles that might, if involved in a fire, endanger the building in question can be most helpful to the inspector in assessing the overall fire safety of the facility. The inspector should also take into account the construction of adjoining buildings, the relationship of windows and other openings between the two structures, the existence of protection such as automatic sprinkler systems within the exposure, and the height and area of the structure being inspected. There is a possibility that corrective action in the form of outside fixed protection, wire-glass windows, or other safety features might be recommended in the fire prevention report as a result of this review of conditions outside the building.

Potential Fire Hazards

The fire safety inspector should check for the existence of combustibles that, if ignited, could cause the spread of fire. No blanket procedure can be recommended in this connection because of variables that must be considered. Hazards that present a danger of fire include flammable and combustible liquids and gases, oxidizing agents, explosives, acids, materials subject to spontaneous heating, combustibles, and explosive dust, as well as ordinary combustible materials in which rapid fire spread might be anticipated. The last category includes excelsior, cotton batting, paper, and other finely divided material that might cause rapid fire spread. The inspector must realize that all of these hazards may exist as a necessary and integral part of the operation of given occupancies.

The fire inspector may expect to reduce the number of hazards in some locations; however, in others the answer may involve providing automatic sprinkler protection, separating fire hazard areas to reduce fire spread, or other steps designed to reduce the possibility of a serious fire developing. But the inspector must be resigned to the fact that it is impossible for most occupancies to operate without having at least some potential hazards on the premises.

The fire inspector must be able to evaluate problems created by combustibles in light of the potential fire spread. Certainly, a clothing factory, for example, cannot be told to eliminate all combustible materials.

However, the extent of storage of combustibles in any location within the factory is directly related to protection available, construction, and possibilities for ignition of materials. The trained inspector must use judgment in deciding on appropriate action in each inspection. Combustibles are essential to the operation of many industries and mercantile establishments.

Potential Causes of Fire

During the course of an inspection, the inspector should observe sources of ignition that could be a cause of fire and take steps to reduce this danger to a minimum. Again it must be realized that it is almost impossible for the fire safety inspector to effect safeguards that would completely eliminate possible fire causes.

Sources of ignition that are frequently responsible for fires include electrical arcs and sparks; lightning; static electricity; friction through grinding, polishing, cutting, and drilling operations; various types of chemical reactions and open flames; improper heating devices; smoking; use of torches; and many other causes. Sources of ignition are present in most occupancies.

It is very difficult, if not almost impossible, to eliminate all potential causes in most occupancies. Certainly, an inspector should exert every effort to reduce fire causes in a structure to an absolute minimum. Conditions can change very rapidly; in fact, it is entirely possible that "cause" conditions abated at one moment are back again almost as soon as the inspector has left the building. As an example, a fire inspector may admonish an individual for smoking in a hazardous area only to have the person light up again as soon as the inspector is out of sight. Fire prevention inspection and control should therefore emphasize the need to restrict ignition sources to actual operating requirements, provide safeguards in installation and handling, and clear surrounding areas of combustible materials in which fire could spread.

Potential Spread of Fire

During the inspection tour, it is quite important for the inspector to consider the ways in which fire might travel within the building. Structural factors that might contribute to the spread of fire are the location and condition of elevator shafts, stair towers, light wells, laundry and trash chutes, pipe chases, conduit openings, vertical and horizontal ducts, dumbwaiters, combustible ceiling tile, combustible interior finishes, windows that might permit vertical spread of fire outside the building, shafts and any other unprotected openings. Access to adequate means of egress is a major consideration.

Major causes of horizontal fire spread are absence of fire division walls and partitions in basic construction or lack of or improper fire doors in existing fire walls. Combined with these structural factors, the presence of the combustibles and hazardous materials, poor stock storage methods, and lax housekeeping practices contribute to the spread of fire. The inspector with a fire suppression background is in a much better position to accurately evaluate fire spread factors than is one who is not familiar with fire fighting procedures and fire spread characteristics.

To make a proper fire safety inspection of the aforementioned points requires a thorough knowledge of building construction as well as an understanding of the travel characteristics of fire. Many fire inspectors unfortunately do not consider possible means of fire spread within a structure; their trip through the premises is primarily for the purpose of noting observable conditions that might cause a fire, such as the presence of combustibles, careless smoking, locked exits, and careless handling of flammable liquids. The mark of a trained, competent fire inspector is an ability to comprehend fire spread possibilities inherent in the design and construction of the building.

Water Supplies and Extinguishing Systems

In touring the building, the fire inspector should check on water supplies for fire suppression purposes and observe standpipe connections, sprinkler systems, and any other fixed extinguishing systems.

In checking sprinkler and standpipe systems, every effort should be made to determine that the systems are functioning properly. Although many inspectors do not consider this a part of the inspection function, it should be a mandatory requirement of every fire safety survey. The fire inspector should not open the valve or otherwise test the device but should request the building operator or owner's representative to do so. Liability is assumed in actually operating a valve, and the facility operator should assume this obligation. The operator has a responsibility to satisfy the fire department that all fire protection devices on the property are operable. One major fire insurance organization believes that if its inspectors check only to see that sprinkler systems are in service, well over 50 percent of their desired results have been accomplished.

Correcting Violations during Inspection _____

As many items as possible should be corrected while the inspector is on the premises. It does not make sense for an inspector to write up items pertaining to blocked fire doors without getting corrective action while on the premises. It is hard to justify not doing so when violations can be easily and rapidly corrected by the person accompanying the inspector. Another example is that of permitting a sprinkler system valve to remain in a shut position rather than having it opened. However, there may be times when the system valve is shut for some valid reason, but this should be checked out and verified and provision made to return the system to operating condition as soon as possible. In the written report, the fire safety inspector should mention items that are corrected during the inspection as a reminder to management to avoid repetition.

Discussing Findings with Escorts

In some cases it is desirable for the inspector to discuss specific problems as they are discovered on the tour. In other cases it would be best for the inspector to make notes of findings and then discuss the locations at the completion of the inspection. The major factor for consideration is the position in the firm of the individual or individuals going around with the inspector. There is no reason to go into a detailed explanation of specific locations of violations with a person who is not responsible for compliance, such as an usher in a theater. However, if the safety director for an industrial plant was accompanying the inspector, it would be desirable to point out the violations during the tour.

The Exit Interview

If someone other than the owner or property manager has accompanied the inspector on the inspection tour, arrangements should be made for an exit interview with a responsible management representative to review and discuss the findings of the inspection. The person in authority may not be immediately available for an interview, and the inspector may need to come back on another day. In any event, the exit interview is an important part of the inspection and should not be overlooked.

During the interview, the inspector should advise the management representative of all the problems noted during his or her visit. If it is necessary to return to fire headquarters for further technical information, arrangements should be made to return or to telephone to discuss these matters.

It is unwise to impose oral requirements that are not later substantiated in the written report. The property owner may make changes based on orally imposed requirements only to find that the inspector, after returning to the fire station, decided they were not necessary. In one case an inspector imposed a requirement for converting windows adjacent to an outside fire escape at a nursing home to wired-glass windows. When the written report on the inspection was submitted, no such requirement was imposed. However, in the several weeks intervening between the inspection and the submission of the written report, the conscientious nursing home operator made the changes at considerable expense. Needless to say, the operator was quite disturbed.

The lesson here is that if on return to headquarters the inspector finds justification for making any substantial changes in recommendations after they have been reviewed in the exit interview with management, the changes should immediately be communicated to the person interviewed, either by telephone or personal visit.

Requirements imposed by the fire inspector should be so clear and concise that they adequately portray to an untrained person the true picture of what is needed for compliance. In explaining the conditions during the exit interview, the inspector should give the owner or property operator the opportunity to offer suggestions on corrective actions. Compliance with requirements will often be more readily attained if the owner or property manager comes up with the solutions to the problems.

Technically, it is sufficient for the inspector merely to state that certain things must be done because the code requires it under a given paragraph. This does not give the property management much satisfaction or understanding of the real need to comply with the fire code for life safety and protection of the property. By explaining as clearly as possible the reasons behind provisions of the code and how they apply, the inspector will make code enforcement much more palatable to management. It may not be necessary to go into great detail on all points, but at least a summary explanation should be given. The inspector must refrain from conveying a threatening attitude about the necessity of fire code compliance. The inspector who settles for "I don't know why but the code says it has to be done" leaves the impression of being not well trained or not sold on the importance of the inspection.

The property owner may be required to spend a great deal of money to comply with fire inspection requirements. The fire inspector has a profound obligation to be sure that all requirements will remedy the situation and not call for unnecessary expenditures. One foolish or impractical recommendation can offset the justifiable aims of the inspection. Management that is not pleased with the inspection will take full advantage of any improper or impractical requirement in attempting to downgrade the entire report. The inspector should clearly identify items that are merely recommendations as opposed to those that are mandatory.

The manner in which the exit interview is conducted has important public relations aspects for the cause of fire prevention, and a knowledgeable and friendly presentation of the findings can go a long way toward making management receptive to compliance with the code.

Report of Inspection

A report of all inspections for transmittal to the property owner or occupant is necessary. In some cases the report is handwritten by the inspector and left on the premises. This practice is not recommended for larger facilities, because it forces the inspector to state in writing all of the problems on the spot with no opportunity to return to the office to look up detailed requirements or possible alternate solutions. In some jurisdictions a completed inspection form is left at the premises.

A report in letter form gives more authority to the recommendations and makes a better impression from the standpoint of public relations. This depends on having report preparation resources available in the fire department. A poorly typed report might convey the impression of an unprofessional inspection.

The inspector has good reason to return to the office to refer to documents for technical information, to give more thought to the requirements specified, or to confer with supervisory personnel on the more complex problems. One disadvantage, of course, is the greater time lag between the completion of the inspection and the time the property owner receives instructions for compliance with inspection requirements. There is always the possibility that a fire might occur during this intervening period; however, the probability of this happening is somewhat remote, and it would seem worthwhile to take the time to prepare a report, with copies of all correspondence to be filed for reference for subsequent inspections or possible legal action.

It is extremely unwise to make a statement or to include in a report words to the effect that the building is free of hazards. There are many potential fire conditions that could bring about a serious fire but that even the most competent fire inspector could not be aware of because they are concealed behind walls, in enclosures, and in other locations where observation is impossible.

Some larger fire departments have fire prevention bureau personnel who are specialized in certain phases of fire prevention work—for example, specialists in flammable liquids and gases, electrical, and other fields. The fire inspector on general assignment may consult with a specialized inspector in preparing the report, or a review of the inspection report may be made by a supervisor.

Reinspection and Procedures to Enforce Compliance _____

Normal inspection procedures followed by most municipal fire departments include a reinspection to ascertain compliance with the fire inspection report. An approximate date for the reinspection should be indicated in the report to notify the property owner when it may be expected.

Procedures to enforce compliance with the fire code require the submission of a report of violations and recommendations for corrective action, as described in the preceding pages. The report usually includes mention of a specific number of days within which the work must be completed. Usually, the dates for anticipated completion take into account the complexity of the job, availability of materials, and other related factors. Housekeeping violations, for example, would normally be given a very short time to correct, while the installation of a stair tower would be given an extended time. A reinspection is made at the end of that time to determine whether the work is underway or has been completed. In some cities, the reinspection is considered the beginning inspection for any legal action taken to enforce compliance. If corrections have not been made, the property owner is given a formal legal order to complete the work within a given number of days or face a possible fine or jail sentence. Normal procedure includes obtaining a warrant for the arrest of the individual who has not completed the corrective requirements within the time specified in the legal order.

Classification of Hazards _____

In the category of hazards, a breakdown between those referred to as *common* fire hazards and those referred to as *special* fire hazards is usually made. The common hazards are those found in practically every occupancy, such as those related to normal use of electricity and heating equipment. Smoking is included in this category. Special hazards are those that the fire inspector associates with conditions unique to a given occupancy.

Hazards in Various Types of Occupancies _____

In reviewing occupancies subject to inspection, particular hazards and causes characteristic of each may be noted. Familiarity with these hazards and causes is helpful to the fire safety inspector in knowing what to look for in each type of occupancy.

Institutions

Long considered the highest risk occupancy from a life safety stand-point, the institution may present a number of unusual situations. A hospital, for example, will contain flammable liquids and gases as well as moderate quantities of combustibles in storage areas. From a life safety standpoint, the institution presents a major evacuation problem in that a majority of the occupants are not usually able to respond unaided to a fire alarm. Personnel within the facility, aided at a later time by fire department personnel, will need to evacuate such a structure or move patients to a place of safety. Common hazards found in such facilities include mattresses, bed clothing, and other combustibles normally found in patients' rooms. The clothing worn by the patient may become a hazard if it is not made of a fire-retardant material. The possibility of a fire occurring as a result of careless smoking becomes greater in a hospital or nursing home because occupants may be immobile.

Other potential fire causes likewise become magnified under institutional conditions. The use of a small amount of flammable gas, necessary in certain medical procedures, becomes a major hazard around electrical appliances. The possible results of an explosion and fire in an institution may be much greater than in an industrial occupancy.

Places of Public Assembly

Another classification of occupancy generally associated with life safety problems is the place of public assembly. Large concentrations of people and the possibility of panic make fire code inspection of this type of occupancy especially important. Conditions in a large public auditorium are quite different from those in an amusement park or a restaurant.

A small fire in the stage area of an auditorium that would have little chance of spreading could cause panic and large loss of life under conditions of improper egress facilities, improper lighting, or overcrowding. The fire inspector conducting a routine inspection may visit a place of public assembly during a time of very limited occupancy. It may be difficult to visualize conditions that would exist if the place were filled with people. Public assembly occupancies also include art galleries, grandstands, tents, libraries, exhibition buildings, funeral homes, churches, skating rinks, transportation terminals, dance halls, and theaters.

Large civic centers and arenas capable of seating thousands of people for special events present special problems. Hazards and causes vary considerably in this broad occupancy classification. However, the one overriding concern is that of assurance of means of immediate egress in the event of fire or other emergency. The possibility of panic always exists, because there is no way of predicting the reaction on the part of an audience. This possibility again compels the fire inspector to reduce fire hazards and causes to the minimum.

In one major incident a small amount of liquefied petroleum gas, estimated at approximately three pounds, was responsible for taking 75 lives in the Indianapolis Coliseum explosion in 1963. In this case, the gas cylinder was located in a storage room adjacent to a refreshment stand inside the structure. Gas apparently escaped and was ignited as a result of a heater for popcorn that was located nearby. The explosion in the confined storage area ripped out heavy concrete and wreaked havoc in the building. The storage of liquefied petroleum gas could have been acceptable from a fire safety standpoint at some other location, but not in a public assembly occupancy. This structure was protected by a well-equipped fire department, and structural deficiencies did not contribute to the disaster.

In public assembly occupancies, fire safety inspectors have usually been successful in getting corrective action to eliminate major fire problems. Constantly recurring conditions that increase the likelihood of problems are overcrowding and locking exit doors. The tendency of some groups to attempt to enter facilities of this type without proper authorization may bring about severe overcrowding and may result in locking of exits to preclude illegal entry.

In restaurants major fire hazards result from deep-fat frying and spattered grease. Protection systems are available for hoods and ducts; however, many older restaurants have not installed such protection. Lethargy in cleanup procedures can result in a hazardous accumulation of grease, which provides immediate fuel if ignited.

Mercantile Occupancies

Mercantile occupancies usually contain many fire hazards and causes. The trend toward large undivided areas for product merchandising has increased the possibility of major fire losses in retail stores. There is also the possibility of delay in notifying the fire department because a fire may not be readily detected in a large area.

Common hazards increase in number in such merchandising outlets, and the counters may contribute a great deal of fuel to any fire. Fortunately, a majority of large retail stores, generally those in excess of 15,000 square feet, are required to install automatic sprinkler protection at the time of construction. In considering the overall potential loss from a fire, the fire inspector in the mercantile establishment should consider water damage. Because merchandise that has been damaged by water usually cannot be sold, the inspector might point out measures that could be taken to reduce the possibility of such damage. This kind of advice may not be required under the code; however, it can be quite helpful in reducing the loss in the event of a fire. It may be feasible for the store operator to place merchandise in the storeroom on skids or pallets to reduce water damage.

Flammable liquids and gases may also be found in mercantile occupancies. Common causes such as defective heating equipment and smoking may be factors in fires in such occupancies. Retail stores may have special hazards associated with the occupancy. An example is a hardware store, where there is a good possibility of explosives being present. Another example is a store where charcoal briquettes are sold. This material is subject to spontaneous heating.

Storage Occupancies

Closely associated with the mercantile occupancy is the warehouse. If material that is stored poses a special hazard, such as large quantities of combustible fibers, additional fire problems are created. Normally, warehouses and other storage occupancies are subject to in-and-out human occupancy. Therefore, there is not as much chance of a fire being started by human carelessness. However, this advantage may be offset by the possibility of delayed discovery because of limited occupancy.

Problems of water damage become even more acute in the warehouse, and serious consideration should be given to provisions that will preclude water damage if sprinklers or hose lines are operated. This may not prevent all water damage, but at least damage in areas away from the immediate fire area will be abated. Outdoor storage facilities such as lumber yards and coal yards serve the same purpose as warehouses. These facilities may be subject to fires resulting in large losses and may even be more readily ignited because of their accessibility.

Office Buildings

Office buildings constitute yet another type of occupancy. Common hazards are usually fairly easy to find in these structures. Common causes are also quite prevalent. A major problem in this class of occupancy is the tendency to overlook combustibles. Personnel occupying an office, as well as even some fire inspectors, may ignore the fuel potential in the paper accumulations in office buildings.

The office occupancy also has a higher degree of life hazard than is usually recognized. Although the density of population is not as great as in a public assembly occupancy, in many large office buildings the density of occupancy is fairly close to that found in a school or hotel.

Consideration must also be given to the inclusion of other types of occupancies in a building that is primarily an office building. A restaurant and its inherent fire problems, for example, may present a life and fire hazard in a structure that would otherwise be considerably safer.

Residential Occupancies

The residential occupancy is the type in which the largest number of fire fatalities occur. Generally, municipal fire department inspectors are not responsible for legal inspection of one- and two-family private dwellings or within individual apartments, although many cities provide for such inspections under a voluntary program. The fire department inspector is responsible for the inspection of public areas of residential occupancies, such as apartment houses, hotels, motels, dormitories, and rooming houses. The inspector is responsible for making sure that people occupying such structures are housed under safe conditions.

Hazards normally found in publicly occupied facilities are usually common ones. Likewise, fire causes generally fall into the same classification. The structures, by nature of their operation, contain combustible materials in the form of room furnishings. Causes are present in the form of smoking material, heating appliances, lights, television sets, and other electrical and heating appliances. It is impractical to eliminate these hazards and causes, although certainly electrical devices can be held to a safe limit.

Legal and Moral Responsibilities of the Inspector _____

The inspector has a legal responsibility to check for all items contained in the code the agency is enforcing. Beyond this, however, is a moral responsibility to point out additional hazards or problems the inspector may note that are not covered under specific points of the applicable code. An example would be an item not specifically required for compliance because of a grandfather clause in the municipal charter or fire prevention code ordinance. The inspector may personally believe that conditions are dangerous, but because of the aforesaid provisions no retroactive application can be required. The inspector has at least a moral obligation to advise the owner or occupant of such conditions.

Subsequent to a major U.S. fire in which almost 100 people were killed, an examination of the most recent inspection report of the local fire department indicated that there were no violations in the facility. The inspection had been made three months prior to this tragic fire. At the time of that inspection, the inspector had observed conditions that eventually contributed to a major loss of life but failed to note them on the inspection report because of the existence of a grandfather clause. The inspector's attitude was that the problems were exempted by the grandfather clause, so it was not necessary to report them.

Grand jury investigations of this fire made public the fact that the building had received a clean fire safety report only a few months before the fire, yet the conditions that contributed to loss of life existed at the time of the inspection.

An interested citizen wrote a letter to the editor of the newspaper in the city after the grand jury findings had become public. The writer stated that she thought all municipal funds appropriated for fire prevention inspection purposes were "down the drain" if the inspection of this building in which the tragic fire occurred was an example of inspection procedures in the community. She made the point that if the inspections were so weak that no problems were reported in this building, the public could have little confidence in a fire prevention inspection conducted by the department. Although the inspector in question fully complied with all requirements imposed by inspection procedures, mention of hazards not covered by the code would have been helpful. Some means should be provided for this information to get into the hands of the facility owner or operator.

The occupant of the facility that has been inspected may feel that it is completely safe from a fire protection standpoint when the inspector leaves the premises or when an inspection report that lists all deficiencies has been corrected. Generally, the inspector is looked on as an individual capable of providing complete and accurate information on fire spread possibilities, possible points of origin of a fire, and other related information.

Although many larger industries and institutions are inspected by representatives of fire insurance carriers, the fire service should not rely on these agencies to shoulder the responsibility of fire prevention. Inspections should be conducted and reported as if there were no other individuals having any responsibilities for making inspections of the place. The fire department inspector is probably the only one entering the premises who has a legal responsibility for conducting an inspection and taking steps to ensure compliance with code requirements. All other fire safety inspections are advisory in nature, because an insurance representative can speak only for the insurance company in determining the force that can be used in bringing about compliance. It is possible for the insured to seek insurance coverage from some other carrier if the inspection requirements are deemed too severe. As another alternative, the insured may elect to pay higher premiums to keep coverage in effect without making changes recommended by the insurance inspector.

Conflicts with Other Agency Inspections _____

In both the exit interview and in the preparation of a report, it must be remembered that the property owner or operator is probably receiving fire-related recommendations from sources other than the fire prevention bureau. It is thus possible for conflicting recommendations or requirements to arise for the same deficiencies. The fire insurance representative might recommend one solution to a problem, while the fire department might recommend another. It is possible that the solution offered by the insurance inspector is as good as or even better than that developed by the fire department inspector. The latter individual should be willing to listen to other solutions to problems as developed by inspectors from insurance companies or by management itself. The inspector should not have a closed mind to solutions offered by others just because it might seem that to do so would be a loss of face.

The fire prevention bureau inspector certainly has the final authority, although some other individuals inspecting facilities also may have legal authority in the fire prevention field. This includes inspectors from the U.S. Department of Labor who are responsible for enforcing the Occupational Safety and Health Act of 1970. In some states, state labor department inspectors have legal responsibility in this area. When management receives two conflicting legally constituted fire safety inspection reports, a difficult decision must be made. In such a situation, management might call a conference of all interested parties as a method for resolving such a conflict.

Sometimes an inspector is discouraged at not receiving a warm welcome on arrival to make an inspection or from a lack of cooperation during the exit interview. This may not necessarily be attributed to a resentment of fire prevention inspectors. The cool reception may be blamed on the multitude of inspections to which the average industrial plant, mercantile establishment, or institution is subjected. In addition to fire safety inspections from public agencies, fire department prefire planning surveys, and sanitation inspections, the average mercantile establishment is also subjected to inspections for occupational safety, fire and casualty insurance, as well as other types of inspections. As many as 15 or 20 different inspection agencies, public and private, may call on a large industrial plant during the course of a year.

Occupancy Self-Inspection

In some occupancies, inspections may be carried out by inspectors employed on a full-time basis in the plant, institution, or mercantile establishment or by individuals stationed at the headquarters office. In the latter case, the inspector from the management level travels from plant to plant to conduct fire safety inspections. These inspections are carried out in much the same manner as those of the public fire service inspector, as the employee has the responsibility of assuring complete compliance with company fire safety regulations. They usually have a thorough knowledge of the business or industry involved, and, in fact, their specific knowledge of company processes often exceeds that of the public fire service inspector, who must try to learn about all types of occupancies.

Some fire departments use self-inspections for a selected, low-hazard group of occupancies. Arvada, Colorado, a district of 150,000 population, has designated certain occupancies for this concept. By the use of computers, the department eliminated those occupancies with high risks, hazardous materials, high exposures, fixed fire protection systems, and places of assembly. They then evaluated the remaining occupancies considering square footage and past fire safety code compliance problems. These reviews identified small businesses and light hazard warehouses as being the principal occupancies suitable for the self-inspection system.

A cover letter and legally correct self-inspection form is sent annually to the selected occupancies. A high percentage is returned promptly. Spot checks are made to ensure compliance.

Potential legal ramifications must be considered before implementing a self-inspection program. The issue of liability for compliance failures due, for example, to misunderstanding the form, is an important factor, as is the issue of the jurisdiction providing fire safety inspection services to some businesses and not to others.

Consultant Fire Safety Inspection

Some mercantile chains and larger industries engage the services of an outside consultant firm to carry out fire prevention duties. These consultant firms assign personnel who are familiar with the given type of industry and have a considerable degree of familiarity with the processes or procedures of the organizations being inspected. If properly supported by the employing company, this type of program can be most effective.

The consultant concept is employed by a number of major concerns that are self-insurers. To be successful, a self-insurance program must include a provision for stern enforcement of fire protection requirements, either by company employees or through the use of an outside engineering firm.

Insurance Carrier Inspection

The average mercantile establishment, institution, or industry is also faced with the possibility of an inspection by a representative of the insurance carrier. This inspection may be for the purpose of determining the insurance rates to apply to the property. Inspections may be conducted to determine possible limits of policy coverage and to tailor the policy to the specific situation encountered.

An insurance inspection may also be carried out for the purpose of rendering engineering services. This inspection may be by a representative of an individual insurance company or by a representative of an insurance group, such as the Industrial Risk Insurers or Factory Mutual System. The primary purposes of such an engineering inspection are to ensure that the fire protection equipment in the plant is in proper working condition and to identify any major hazards. Changes in procedure, layouts, construction, and protection are also noted.

The property owner will find that the inspection for engineering purposes includes testing fire protection equipment and available water supplies. This could include the tripping of dry pipe valves and flow tests on hydrants, tests of stationary fire pumps, and tests to ensure proper operation of all fire protection equipment in the structure. Although these inspections are most helpful, they are an interruption to the routine of the plant. Management must provide personnel to carry out the duties of opening valves, resetting dry valves, etc., as a part of the tests.

Other Government Agency Inspections

In addition to the previously mentioned private inspections, the industrial plant or mercantile establishment will from time to time be inspected by other governmental agencies. The health department, for example, will probably inspect the facility for overall sanitation and for food handling procedures. Institutions will likewise be inspected by the health department to determine compliance with standards relating to nursing care, patient health, administering drugs, etc.

The labor department or department of labor and industry in practically every state has a responsibility for employee safety. These inspections may be conducted by the U.S. Department of Labor inspector. As mentioned in Chapter 10, this responsibility may cover the provisions of the Occupational Safety and Health Act of 1970. The safety inspector will usually take immediate action on conditions encountered that affect the safety and well-being of the employees, including basic fire prevention items.

The local industrial plant may have prefire planning inspections by personnel from fire fighting companies separate from fire prevention inspection activities. Regardless of coordination and timing, personnel from the plant should take the time to accompany the inspectors in carrying out these duties.

The federal government may send inspectors into given plants through the provisions of the Port Security and Waterways Act of 1972. This act provides for Coast Guard inspection, regulation, and enforcement in facilities adjacent to navigable waterways. The Coast Guard is empowered to bring about compliance, which includes fire safety, in such facilities.

A federal inspector may likewise inspect a building for fire protection purposes if the plant has contracts for providing services or materials to federal agencies or has processes subject to federal inspection, such as food preparation. Such inspections are designed to ensure the continued or safe production of items under review and may include coverage of fire safety items. In some plants, an inspector may be kept on the premises at all times.

Although all of the preceding inspections are necessary, they consume a considerable amount of management's time as well as that of maintenance personnel. Plant management may not welcome the fire inspector because of the multitude of inspections. Time taken by plant personnel in accompanying the inspector and by management in discussing requirements is time away from production.

Summary

A key function in the enforcement of fire laws and regulations under the jurisdiction of the fire department is the fire safety inspection. Beside having a thorough knowledge of code requirements and competency in inspection procedures, the fire safety inspector should have good judgment, keen observation, and skill in dealing with people.

Before going out on an inspection, the inspector should be prepared with the proper tools: a flashlight, camera, notebook, recorder, or computer, for taking notes on the inspection tour, and code books as well as handbooks available in the car for reference. The inspector should also look over any reports of previous inspections of the facility for background information to enable proper follow-up on earlier recommendations. Under no conditions should an inspection be undertaken when the inspector is pressed for time and unable to make a thorough inspection.

On arrival at the site, identification must be presented to a person in authority, and permission must be secured to make the inspection. This is a legal obligation as well as a courtesy. Court cases have challenged the legality of inspections made without proper identification and permission. It is customary also for the inspector to be accompanied by a representative of management.

The tour usually starts at the top of the structure with an inspection of the roof, then proceeds to the top floor and on down through the entire structure, including the basement levels. A systematic inspection should be made of all spaces on every floor; if access to certain spaces is not immediately possible, the inspector must wait until permission for entry to these can be arranged.

The objective of a fire prevention inspection is to discover hazards and make recommendations or impose requirements to abate or eliminate them to the greatest extent possible. Conditions to look for are exposure hazards, potential fire hazards outside and inside of the building, potential fire causes, potential spread of fire, water supply, and extinguishing systems. The inspector should think about everything observed on the tour and make a point of registering the significance of hazards, exit deficiencies, or any unusual conditions. Beside noting the obvious potential fire causes or storage of flammable liquids or gases, the competent inspector should look for latent hazards, such as ignition sources from operating equipment, and should be able to recognize fire spread possibilities inherent in the design and construction of the building. Under no conditions should the fire inspector leave the premises without being sure that all fixed fire protection equipment is in proper functioning order as required by applicable fire prevention codes. And aside from the mandatory responsibility to ensure compliance with legal requirements of the code, there is a moral obligation to point out to property owners any potential hazards not specifically named in the code.

As many easily correctable violations as possible should be taken care of during the inspector's visit. Changes necessary for compliance with the fire code should be discussed with management at the end of the inspection tour. Recommendations should be clear and concise and should be carefully weighed to make sure they will remedy the situation and not put the owner to needless expense.

A report should be made of all fire prevention inspections to give the owner notice of violations and to have a record on file for follow-up of recommended corrections and for reference in possible legal action to enforce compliance. An owner is usually given a specific time in which to complete the requirements; if reinspection finds the work not completed, a legal order is issued to comply or face a possible fine or jail sentence.

Among general considerations, the fire safety inspector should be aware of hazards peculiar to various types of occupancies, such as places of public assembly, institutions, mercantile establishments, warehouses, office buildings, hotels and residential occupancies. For example, such hazards include blocked exits or passageways in public assembly places; grease in restaurants; the hazard of fire in institutions with bed patients or elderly residents. The inspector should also be aware of possible conflicts with inspections carried out by insurance carriers and other local, state, and federal agencies and with industry self-inspections and municipal grading schedule surveys. Because the various inspections a property is subject to may put an owner in a quandary, the fire inspector should be prepared to cooperate in working toward an accommodation of all interests without mitigating fire safety.

Note

1. Amendment 107, *Volume V Fire Prevention Code* (Raleigh, N.C.: North Carolina Building Code Council, 1991).

CHAPTER 6

Enforcement of Fire Safety Codes

Foremost among the responsibilities of the fire prevention bureau is the enforcement of the municipal fire code. Some of the approaches that have proven to be effective in attaining conformance to fire safety codes and regulations are examined in this chapter. Inspection is an intrinsic part of most of these approaches, but because the subject covers so much ground and deserves special attention, it is treated separately.

Publicity for Fire Codes

Publicity regarding fire safety codes and the enforcement of such codes has not been used as often as it could be to educate the public in fire safety. As one example of effective use, the fire prevention bureau should seek press coverage in connection with hearings for enactment of fire prevention codes. Although this may bring out individuals who are opponents of the legislation, publicity gives citizens a clearer idea of fire code coverage. Media publicity on code enforcement procedures is an excellent way to reach the public. Press publicity of code enactment is too often limited to occasions when a requirement is imposed to abate a specific hazard that has come to light as the result of a tragedy.

Signs placed in public buildings indicating that certain activities are violations of the fire safety code such as those indicating "No Smoking" or giving occupancy capacities, such as "Maximum Capacity, 1,000— City Fire Prevention Code" remind citizens of fire safety.

Plan-Review Program for Fire Code Enforcement

A number of municipal fire departments have programs for the review of plans and specifications for buildings to be constructed within their jurisdiction. Qualified individuals are assigned to check plans and specifications for conformance to fire code requirements.

Usually, these reviews are coordinated with those carried out by the local building department. Plans are received originally by the building department and are then routed through the fire department as a matter of procedure.

A plan-review program can best be established through a municipal directive or state law stating that all plans for construction be reviewed by the fire department's fire prevention bureau.

The scope of plan reviews carried out under these arrangements is not always as broad as it might be. In a number of jurisdictions, the review is confined to checking the locations of fire hydrants, accessibility for apparatus, and of standpipe connections. Major items of concern in fire protection, such as exitways, interior finishes, basic construction, and sprinkler layouts, often are not checked by the fire department.

Personnel need professional training to carry out these responsibilities. They cannot be expected to participate in detailed reviews of plans and specifications and in conferences with architects, engineers, and builders without having an adequate background. Fire service personnel who have been assigned to plan-review duties without proper training and background have caused a loss of prestige professionally for the departments they represent.

Fire Protection Engineers for Plan Reviews

Ideally, a fire department will have a fire protection engineer available to carry out plan reviews, reviews of specifications, and consultations with architects, engineers, and builders. The desirability of having professional personnel who can speak with authority in the engineering field assigned to plan-review duties cannot be overemphasized. Smaller municipalities may find the addition of a fire protection engineer to the payroll prohibitive in cost. Chicago, Los Angeles, and Austin, Texas, are among the cities with fire protection engineers assigned to the fire department. The fire protection engineer is usually in a civilian, nonuniformed force and in a staff rather than a line position.

In addition to having duties connected with review of plans and specifications, fire protection engineers are frequently concerned with long-range departmental planning relating to station locations, response districts, etc. They may also have considerable impact in the development of fire prevention codes and regulations. In view of their assignment in staff positions, they are usually not responsible for directing inspectional programs and other line functions.

In smaller municipalities, where the possibility of employing a fire protection engineer may be remote, the service might be more logically provided at the state level. The state fire marshal's office might have a fire protection engineer on staff who could be made available to assist with plan reviews in jurisdictions that find it difficult to provide an engineer of their own. This arrangement has an advantage from the standpoint of work load and is definitely a procedure by which best advantage can be taken of available talent. Another advantage to having the fire protection engineer at the state level conduct the plan review is that the review would encompass consideration of both state and local fire safety codes, as well as consideration of the sections of the building code that pertain to fire protection. Some small jurisdictions contract with nearby cities to provide this service on an "as-needed" basis.

Plan-review programs should, if at all possible, provide for review of preliminary plans. Some jurisdictions do not permit personnel to take the time to review plans in the preliminary stage; however, review at that point is usually rewarding. The entire direction of a project from a fire protection standpoint can often be fairly easily changed at the time of a preliminary review; however, a change at a later stage is usually considerably more difficult and many times more costly. Public relations are also improved by an early review of plans, specifically in the preliminary stage. Fire departments that do not conduct reviews of preliminary plans undoubtedly end up spending a greater length of time reviewing final plans than would have been necessary had a preliminary review been conducted, and problems that might have been easily corrected if detected in a preliminary review can develop into major problems at the time of a final review.

Most architects and engineers appreciate having a plan-review service available. It is considered a great value to these professionals in the performance of their work, because a plan review serves as a check procedure. Plans should be submitted in such form as to permit detailed review.

Although consultations with architects during a plan review are necessary and desirable, the reviewer should have the opportunity to study the plans in detail alone, without interruption or distraction. Notes on the review should be kept on file, and narrative comments should be made to the architect. In some communities, the shortage of personnel has made it necessary merely to mark comments on plans and return them to the architect. This arrangement leaves a great deal to be desired because it permits no retrieval of notes and recommendations once the plans have been returned.

Microfilming of Plans

Some communities microfilm the plans submitted for review. This enables plan-review personnel to retrieve information months or years after the building has been constructed, which is especially desirable where alterations and additions may be made at a later date. Microfilming of plans also has the advantage of reducing storage space requirements. It is practical to retain the plans in their original form until the building has been constructed and then reduce them to microfilm. An ability to retrieve plans is also desirable from a legal standpoint. Questions about details of approved plans may arise many years later, especially after a major fire in the building.

Training of Fire Service Personnel for Plan-Review Functions

A community in which it is not feasible to hire a fire protection engineer, and for which the state or county does not provide such services, can have an effective plan-review program by preparing other personnel for such duties. In most fire departments, at least one person, possibly a fire fighter with drafting experience, could be trained in reading construction drawings and in fire protection requirements. This at least provides some representation of the fire service in the plan-review process. It will be more difficult for such a person to gain acceptance by architects and engineers than for an engineer or fire protection graduate.

Code Backing for Plan Reviews

In reviewing plans, the fire protection engineer or other person carrying out these responsibilities should have a designated code for reference to provide full backing for decisions. As previously mentioned, certain building code requirements might be applicable. If fire safety-related portions of building codes are to be enforced by a fire department review person, prior arrangement must be made with the building official to avoid conflict and misunderstanding.

Plan-Review Correlation with Inspection

Plan reviews are only as effective as the inspection follow-up to ensure full compliance with requirements noted by the reviewer. Requirements set down by letter from a plan reviewer or by red pencil marks on a set of plans may not always be carried out by either the architect or the builder. Unfortunately, omissions cannot always be readily detected in the field because the deficiency may be enclosed during construction and may not be revealed until a fire occurs. Inspections should be scheduled as necessary to coincide with the completion of various stages of construction so that the inspector can be at the site to detect any deviations from specifications.

Cooperation in Plan Reviews

Cooperation among several agencies is absolutely essential for accomplishing the purposes of the plan review. These include the building department, the fire department, state-level agencies, and any other governmental agencies concerned in the plan review. In a number of jurisdictions, all agencies having responsibility for plan review and code enforcement meet periodically to review plans and to discuss problems with architects, contractors, and other interested people. Some jurisdictions also include zoning, planning, traffic control, environmental services, and law enforcement representation.

The insurance industry may also have an interest in reviewing plans for construction, especially in major industrial and institutional facilities. There is a possibility of a conflict between fire protection requirements of the insurance carrier and those of the fire department. Although it is presumed that the fire department's requirements must be met, on occasion arrangements might be worked out to encompass the requirements of the insurance industry as well through alternate procedures.

A requirement made by an insurance carrier reviewer is not necessarily out of order just because it is not found in the fire safety code. Of course, the fire department cannot place itself in a position of officially requiring adherence to provisions other than those for which it has a legal responsibility for enforcement. On occasion, joint plan reviews between insurance and fire services personnel have been successful.

Control of Sales and Use as a Means of Fire Code Enforcement _____

The control of sales and use is another means by which codes for fire abatement and prevention may be enforced. Most fire codes contain provisions that give the fire prevention bureau responsibility for controlling sales of explosives, fireworks, fire extinguishers, fire alarm systems, and other related devices, as well as general control of the handling of sales of gasoline and other hazardous materials.

Control of Gasoline and Other Hazardous Commodities

Control of the sales and use of gasoline is found as a subject in practically all fire prevention codes. Included are required procedures relating to the handling, storage, and dispensing of gasoline, with specific requirements relating to the type of container into which gasoline may be pumped, how gasoline can be dispensed, the prohibition of smoking, and requirements that engines be turned off. The fire prevention bureau is required to enforce these regulations and to see to it that service station operators fully comply with all provisions related thereto.

Control of sales also includes sales, handling, and storage of other dangerous and hazardous commodities. In all cases in which the fire prevention bureau is involved, there is a possibility of ill feeling in that no business person enjoys being told that a product is unsafe, is not listed, or otherwise cannot qualify to be sold within the city. Political pressures may be involved, and the fire prevention official could be threatened with lawsuits in connection with this type of enforcement.

Specific sales methods are also subject to control. An example is self-service dispensing of gasoline. This method of dispensing gasoline has been given a great deal of study by public fire service personnel. Self-service dispensing is very popular. It appeals to the customer because it is quick, and the economic factor is a valid concern of the service station operator. The fire safety officer, of course, should not include economics as a factor in code enforcement.

Control of sales and use of hazardous materials are delegated to the fire department in many jurisdictions. Most fire codes specify which hazardous materials present a fire or explosion danger and include their control as a fire department function. Some jurisdictions assign responsibility for materials that present solely a toxicity hazard to some other agency.

Federal "right to know" provisions mandate that fire departments be notified of hazardous materials in their service area. Although the requirement is primarily to facilitate fire fighting, the data can be helpful to the fire prevention bureau as well.

Some jurisdictions require hazardous materials storage locations to be placarded to assist fire fighters. Generally, the fire prevention bureau is responsible for enforcement of these mandates.

Control of Fire Alarm Systems

Control of the sales of fire alarm systems, including smoke detectors, is a public safety function. A fire alarm system is sold to the public for fire safety purposes and must be properly designed to function in the event of an emergency. From a public safety standpoint, the average person may have no way of knowing whether a fire alarm system, especially one designed to detect fires, is functioning properly until there is an emergency. In the case of practically every other type of household equipment, including toasters, radios, and can openers, the buyer can try the product at home to see whether it will perform satisfactorily. Some communities require that all fire alarm systems in the community be approved by the fire department. Most jurisdictions require the installation of smoke detectors in residences; some only in new homes, whereas others include new and existing residences.

Control of Fire Extinguishers

Fire extinguishers are also regulated at the point of sale in some jurisdictions. A number of inferior, substandard fire extinguishers are on the market. They have extremely limited value in the control of fires; however, they are frequently advertised as having superior fire suppression capabilities. Regulations enforced in some areas are designed to keep inferior extinguishers off the market.

Control of Explosives

The control of handling, storage, and sales of explosives has long been covered in fire prevention codes in hundreds of municipalities, as well as in fire codes of counties and states. Explosive storage is considered in zoning regulations as well. The control of the sales of explosives may include requirements for identifying people who purchase explosives, as well as listing the names and addresses of purchasers. Control in this field is also a function of the federal government through the explosives regulations enforced by the Bureau of Alcohol, Tobacco and Firearms.

At the local and state level, explosives control is considered a responsibility of fire prevention code enforcement agencies. Although explosives are actuated by fire or percussion, the logical avenue for control appears to be within the fire prevention field; for this reason, the fire marshals' offices or fire prevention bureaus have usually been designated as the agencies for enforcing all regulations relating to the control of explosives.

Security Measures. Problems in the field were relatively simple until the wave of protest movements brought widespread use of explosives by militant individuals and groups. Regulations may control the sale of explosives to ensure that they are limited to bona fide purchasers and thus to preclude acquisition of explosives by militants and others with destructive purposes. Theft then becomes a major problem, and secure storage of explosives must be ensured. Regulations for storage magazines usually specify theft-resistant as well as fire-resistant construction and security locking devices; in some instances, security fencing may also be required.

Deactivation Measures. The control of explosives may also require the provision of personnel to deactivate or otherwise render safe clandestine devices. The use of bombs and timed devices incorporating explosives has increased considerably, and control measures must be taken by the community to ensure that personnel are available to render devices harmless.

Several municipal fire departments and state fire marshals' offices have responded to this need by developing bomb squads for the purpose of handling explosive devices. One of the most successful programs is that conducted by the Shreveport, Louisiana, fire department. State fire marshals in several states, including Iowa and Maryland, operate such squads. North Dakota's fire marshal operates a statewide hazardous materials response unit.

Police Jurisdiction. In some communities the explosives control responsibility is under the jurisdiction of the police agencies because an intentional explosion is a criminal act that is under police jurisdiction for investigation. There are some excellent police bomb squads.

Fire Department Jurisdiction. The assignment of responsibilities for explosives control within the fire service has the advantage of a close tie-in between inspection of storage locations and possible theft and misuse. In the event that responsibility for full control is assigned to the fire organization, personnel must be fully conversant in investigation techniques as well as have a knowledge of explosives.

Control of Fireworks. Another field of responsibility normally assigned to the fire service, but which often involves police agencies as well, is that of fireworks control. This control responsibility probably has as many varieties of approach as any subject in the fire prevention field. There are many differences in requirements relating to fireworks from state to state and even within counties in a given state. Perhaps beginning with Nova Scotia's legislation in 1762 to prevent the firing of squibs, rockets, serpents, and other fireworks, laws relating to the use of fireworks have varied widely from jurisdiction to jurisdiction and from year to year.[1]

A number of states do not permit the sale or possession of fireworks except for public displays conducted with the permission of the state fire marshal's office or local governmental agency. In other states certain types of fireworks may legally be sold over the counter. There are no states in which all types of fireworks may legally be sold. Generally, the "closed" states have provisions for fireworks possession only for displays sanctioned by the enforcing agency. Permits for such displays are usually granted only after a bond has been posted or insurance coverage is in effect to cover accidents that may occur during the event. By issuing these permits, the state sanctions an activity that would otherwise be illegal; therefore, special precautions must be taken to ensure safety.

Generally, the issuance of fireworks permits is a function of the fire prevention agency. Sale and possession of illegal fireworks is done in a clandestine manner, which requires detailed investigation to abate. This enforcement activity may be carried out jointly by fire and police personnel.

Fireworks are a danger primarily to personal safety, though they may also present a hazard with respect to fire safety. As in the field of explosives, training is needed to carry out an effective fireworks control program.

Control through Limitation of Sales. Control is usually effected through limitations on sales or requiring permits for possession; however, a major problem always exists when one jurisdiction permitting sales adjoins a "closed" jurisdiction. The public can merely cross the state line or municipal boundary to purchase fireworks and then return home. Public sale of certain types of fireworks is permitted in unincorporated areas of a number of states, but major cities and towns in these states do not permit possession of fireworks without a permit.

Enforcement of local regulations is difficult where items may be purchased nearby. Public education programs have been used in an effort to control this problem. In some cities, fire apparatus has been placed at the city limits with signs affixed to warn the public of the illegality of fireworks within the city. Loudspeakers may also be used.

In several jurisdictions, unmarked cars have been used to trail purchasers from the point of sale to a point within the city in which possession is illegal. A stop is then made, and action is taken to seize the material and arrest the perpetrators.

Legal Ban on Fireworks. Fire service organizations have been in the forefront among agencies attempting to require a legal ban on the public sale or use of fireworks. In fact, fire service organizations are probably primarily responsible for restrictive legislation in the majority of states.

A federal law enacted in 1954 prohibits the transportation of fireworks into any state for use outlawed by that state. Offenders are subject to a maximum fine of $1,000 or a year's imprisonment or both.

Another organization that has had a major interest in fireworks legislation is the National Society to Prevent Blindness. This society is especially interested in eye injuries, but its effective work has not been limited to reducing eye injuries. The organization has successfully worked in a number of states to promote prohibitive legislation. This society works in cooperation with the National Fire Protection Association to enhance public awareness of the dangers associated with use of fireworks.

Public Fireworks Displays. In that public fireworks displays with permits are permitted in practically all jurisdictions, it is incumbent on the fire marshal's office to monitor those events. Provisions should be made for closing down the display in the event of problems. In some jurisdictions, individuals who fire displays are required to be certified by the fire prevention agency.

Enforcement of Regulations. Enforcing explosives and fireworks regulations brings fire prevention bureau personnel into contact with members of the public with whom they usually have little contact. The explosives control field, for example, involves a considerable amount of contact with the construction industry. A close relationship must be maintained with these individuals to ensure safe operations.

Likewise, fireworks control, if effective, may create a number of adverse situations. The fire prevention inspector may be placed in the position of having to seize fireworks and file juvenile petitions against middle school or high school students. Probably most other contacts by the fire prevention official with people in this age group are of a positive nature. The confiscation of fireworks that have been purchased in another jurisdiction is usually not well received and definitely places the enforcement official in an adverse situation. Because of this, there may be a reluctance on the part of fire safety personnel to follow through with full enforcement of the regulations.

Structural Control as a Means of Fire and Building Code Enforcement _____

Structural control of buildings—another means of enforcement—is effected through the previously mentioned plan-review function coupled with inspection of structures, before and on occupancy, to be sure that all fire safety requirements have been fulfilled. This inspection may be conducted in company with building officials. Structural control includes inspection of materials, appurtenances, and other factors relating to the building.

Fixed Manufacturing Equipment

Fixed equipment used in processing and manufacturing is included within the scope of structural inspections for fire code enforcement. Some of this equipment has design features that may lead to fires. These hazards must be ferreted out in the inspection process. On occasion the equipment may increase in degree of danger as it is used. Again, it is the responsibility of the code enforcement person to see to it that such equipment is noted and that proper repairs are made.

Interior Finishes

Control of the structure relates also to interior finishes. This is an important aspect of fire protection that has often been overlooked during fire safety inspections. Proper evaluation requires analysis of materials used in a structure to determine that the materials will not readily burn or that if they are ignited rapid spread of fire will not ensue.

Figure 6-1 Fire suppression companies conduct fire safety and prefire planning surveys in many jurisdictions. (Clark County, Nevada Fire Department.)

Fixed Fire Protection Equipment

Inspection of fixed fire protection equipment is another part of the job of control of structures. Tests and inspections at the time of occupancy are also quite important. As with other phases of fire protection inspection, personnel assigned to this responsibility should be familiar with testing procedures necessary to ensure proper operation.

In some fire departments, fire suppression personnel are called on to assist fire safety personnel with these inspections and tests. Special equipment may be necessary to properly test fire protection devices. Frequently, water supply tests must be conducted at night and on weekends because of the problems created by the flow of water over streets and other public areas. It is always prudent to have the owner's representative perform the test with the inspector as a witness.

Reinspection

Control of structures must be maintained by continuing to inspect buildings to note changes that take place from time to time. The addition of a partition or change in the swing of a door, as insignificant as it may seem, could make a major difference at the time of a fire. All such changes can be noted in a structural control inspection program. Reinspection of buildings should also include inspection and supervision of testing of fire protection equipment on a periodic basis.

Temporary Conditions

Inspection for the control of structures must also recognize temporary conditions that may retard chances for egress in the event of a fire. This refers to inspections to determine that exits are not being blocked by furniture, cabinets, or other appurtenances and to determine that exits are not being locked when the building is occupied. This type of inspection is similar to that carried out for control of occupancy.

Electrical Equipment

Another phase of the inspection procedure relating to the control of structures is that of checking electrical equipment and devices. Although fire safety inspectors do not normally carry out a full range of electrical inspection duties, some elements of electrical safety inspection are noted by the fire inspector on routine inspection. Frayed wiring, obvious overloads, circuits not identified, extension cords, arcing equipment, and inadequate fusing are conditions normally found within the scope of the fire inspector's responsibilities.

Control of Occupancy as a Means of Fire Code Enforcement

Control of occupancy is another enforcement measure. This type of control includes determining and posting occupancy capacities within structures, coupled with the periodic presence of fire prevention bureau personnel in the facility to ensure that an excessive number of people do not occupy it and that all exits are usable.

An additional example of control of occupancy is the assignment of fire department personnel to public events to see that aisles are not blocked and that overcrowding does not occur. These duties are most effectively performed by personnel in uniform. Control of occupancy may also encompass programs designed to make sure that company employees know how to operate fire protection equipment on the premises. Examples are the training of employees of places of assembly in the use of fire extinguishers.

People creating a disturbance in audiences have made the job of occupancy control a more difficult one. On some occasions, overcrowding has been reduced through effective control by fire personnel. It is recognized that the involvement of fire service personnel with an operation of this type usually does not improve public relations. People who are barred from entering premises because of fire code capacity requirements are often angry and may on occasion take direct action against fire department personnel. Many people resent being told what to do in a place of assembly, especially if they have paid to enter the premises. Extreme tact must be used at all times in this extremely important duty.

The fire department's activities in the control of occupancy may include efforts to ensure adequate fire reaction. There is a close relation between fire safety inspection and measures that contribute to life safety through instilling positive fire reactions.

Legal Aspects of Fire Code Enforcement _____

In compliance and abatement procedures for fire safety code enforcement, the ultimate goal is the improvement of the facility from a fire safety standpoint rather than imposition of fines or other punitive measures. There is little satisfaction in having fines levied or jail sentences imposed if in the fact the facility remains unsafe. Court actions are time-consuming and should be considered as a last resort rather than as a primary means of enforcement in fire prevention.

Some fire prevention personnel carry notices that serve as official notification of violations. In some cases inspectors may impose fines when they issue these notices. This procedure enables the fire department to immediately remedy hazardous conditions and is somewhat analogous to the police officer issuing a traffic citation or ticket to an errant motorist. The logic is that a fire code violation is endangering others and the culprit should be dealt with in an equally harsh manner.

In addition to serving legal orders, fire departments may on occasion have reason to obtain injunctions or other sanctions to bring about prompt action to remedy a serious fire hazard. Since procedures vary greatly from state to state, no attempt will be made to delineate specific legal processes. Fire department personnel having enforcement responsibilities should meet with the attorney for their jurisdiction to learn about specific avenues available for enforcement purposes. It is far better to obtain this information in advance of an incident rather than on the spur of the moment when life is in jeopardy in a facility.

Decisions handed down by several state supreme courts have a direct bearing on fire code enforcement. These are in addition to U.S. Supreme Court decisions cited in Chapter 5 with respect to proper identification of inspection personnel and permission to inspect.

The first of three Alaska Supreme Court decisions was in *Adams v. State of Alaska*. This case came as the result of a hotel fire in Anchorage in which five lives were lost and several guests were injured. Representatives of the injured and deceased filed suit for damages, naming, among others, the state as a defendant. The state fire marshal's office, pursuant to legal authority vested in it, had made an inspection of the hotel more than eight months before the fire.

Representatives of the fire marshal's office had found an inoperative fire alarm system and several other fire safety violations. The court noted that the fire marshal's office had not taken positive steps to notify the hotel of the violations. In opposition, the state contended that it could not be held liable for damages.

The Alaska Supreme Court held that the state, by undertaking to inspect and advise on conditions in the hotel, assumed a duty to those injured in the fire. The possible ramifications of the decision, which upheld the right of suit in this case and in which the state was negligent, were addressed by the court, which stated, "We think it unlikely that limited liability for negligence in an inspection will force the state from the field of fire inspection. Fire prevention is a recognized government function, not an experimental program. The cost of fire prevention, including the risk of liability . . . is still less than the cost to the state of disastrous fires, in terms of fire fighting effort, lost taxes and the impact on the economy."[2]

In the second decision, in the case of *State of Alaska v. Jennings,* the court also dealt with a hotel fire.[3] The Nordale Hotel in Fairbanks had a fire that caused the loss of 11 lives. The survivors sued the state, the city of Fairbanks, and the hotel management. In this case the state fire marshal had deferred inspection in Fairbanks to the city fire prevention bureau in accordance with agency policy, as permitted by Alaska statute. The state, the court held, had no responsibility under these provisions.

The city fire prevention bureau had made inspections of the hotel over a period of years and had sent many letters to the facility requiring correction of deficiencies, all of which were ignored by the operators. The court held that the city was negligent in allowing continued operation of the hotel and, further, that the city had an obligation to obtain compliance once inspections were commenced.

In the third case, *Northern Lights Motel, Inc. v. Sweaney,* testimony pertained to an Anchorage motel fire in which there was only one victim.[4] The victim's representative had successfully sued the motel in lower court, and the motel appealed the decision.

The fatal fire started in a room occupied by another guest. It spread into the room in which the victim was staying. Testimony was given that the motel was constructed largely of 3/8-inch to 1/2-inch plasterboard or sheetrock, whereas the prevailing fire safety code required at least a one-hour fire rating, which would necessitate use of at least 5/8-inch plasterboard.

Inspections conducted by the local fire department revealed certain defects in ceiling construction; however, there had not been a determined effort to achieve compliance. The Alaska Supreme Court held that the owner had an obligation to build the structure in accordance with codes and therefore found the suit to be in order. An expert witness had testified that the construction deficiency could well have permitted the fire to spread, which brought about the death.

In a case originating in Des Moines, the Iowa Supreme Court held that the city could be held liable for death and injuries in an apartment house fire. The city had inspected the building before the fire and had issued numerous violation notices but had provided the occupancy with an inspection certificate "which impliedly warranted the premises to be safe for the purposes of human habitation." This case, *Wilson v. Nepstad,* was decided by the Iowa Supreme Court in 1979.[5]

Supreme courts in some other states, however, have not supported the concept that the enforcing authority is liable for failures in assuring compliance with codes. An example is the case *Nolan v. Parker et al.,* heard by the Appeals Court of the Commonwealth of Massachusetts in 1983. In this case an apartment occupant sought damages for personal injuries, loss of wages, and loss of personal property resulting from a fire. The building owner, town government, and others were sued. The town's building inspector had ordered fire safety corrections but had not reinspected to ensure compliance. The fire occurred three weeks after the inspection.

The court held that the town's duty was to the public generally and did not include "special duty" to the plaintiff. The court further noted that "The traditional rule, that a building code is enacted for the benefit of the public and therefore that its violation does not give rise to a private right of action, continues to be followed by the majority of the states that have considered the question." The term *violation* in this case refers to the town's responsibility to fully enforce the code.[6]

Condemnation of Unsafe Structures _____

In addition to activities normally thought of as fire prevention measures, statutes in many states and ordinances in a number of cities provide fire enforcement officials with the power to condemn unsafe buildings. This power is granted to permit the fire official to order the demolition of structures that are found to be hazardous if the owner fails to take corrective action. Generally, the structures must be of combustible construction in order to impose the provisions of the statute or ordinance. It is presumed that a combustible building could endanger other properties if it caught fire.

Most condemnation statutes and ordinances also require that the building be situated in such a manner as to endanger other properties if ignited. In some jurisdictions exposure to a telephone line or other utility line is sufficient to invoke the provisions of the statute.

These ordinances and statutes usually contain a requirement that the structure must also be unoccupied. There is also a requirement that the structure be open for trespass and therefore so situated that individuals can enter the building without difficulty. In some cases arrangements may be made for boarding the building, thereby securing it in such a manner as to preclude the possibility of entry by trespassers, except by force.

The reasoning behind condemnation provisions is that structures open for trespass and of combustible construction that expose other properties contribute to the fire problem within a community. Such structures are open invitations to arsonists, and are fire hazards. Many fires are set by juveniles in such locations.

Most condemnation ordinances contain requirements for the demolition or repair of structures in question, followed by the initiation of legal suit to recover costs. There may also be provisions for the recovery of court costs incurred.

Condemnation procedures have been aided by the willingness of fire departments to use structures for training purposes. Although this demolition method has worked well in a number of communities, it creates a possible conflict with air quality control requirements.

The condemnation of structures as a fire prevention measure is usually welcomed by other property owners in the neighborhood. Most buildings that are candidates for condemnation are unsightly and lower property values in the neighborhood. When the property is owned by an estate, it may be difficult to locate all the heirs to obtain service of the required notice. No work should begin in the demolition of the building until it has been definitely determined that the legal owner has been contacted.

Summary _____

The enforcement of the municipal fire code comes under the jurisdiction of the fire department and is one of the major responsibilities of the fire prevention bureau. Although fire safety inspection is the key function in code enforcement, other forms of enforcement endeavors have proven effective in attaining the objectives of conformance to fire safety codes and regulations. Among these are publicity efforts through newspaper, radio, and television coverage to bring to the attention of the public the issues in code enactment and enforcement.

One way through which the fire prevention bureau can effectively approach enforcement of codes is the plan-review program, in which all plans and specifications for new construction are routed to the bureau to be checked for conformance to fire codes. Inspection follow-up is essential to make sure that fire safety requirements imposed at this stage are carried out in new construction.

Control through limitations of sales is also an effective measure in enforcement. This method applies to sales of gasoline and other hazardous commodities, fire alarms, fire extinguishers, explosives, and fireworks. Structural control is another means of enforcement; it combines the plan-review function with actual inspection of structures, before and on occupancy, including inspection of all materials, appurtenances, and fixed equipment, such as fire protection equipment. Closely related to this function is the prefire planning survey, which is carried out, usually separately from structural control inspection, for the purpose of familiarizing fire service personnel with building locations and conditions in advance of a fire. With this knowledge they can more effectively attack the fire and save time and the lives of fire fighters and occupants.

Control of occupancy as a measure for fire code enforcement entails determining and posting occupancy capacities within a structure and monitoring of places of public assembly by fire service personnel on duty to avoid overcrowding and other violations.

All of these procedures require constant surveillance through a program of inspection and reinspection to make sure that the spirit and the letter of the code are carried out. Fire safety codes are effective only when enforcement efforts are coordinated, fair, and unflinching. Many communities have excellent codes on paper but fail to follow through with effective measures for enforcement.

Other aspects to be considered are the legal implications of code enforcement and condemnation of unsafe structures. Legal measures or court actions to enforce fire safety codes should be considered only as a last resort, though the fire department should be informed of the avenues available. As a legal proceeding, fire code enforcement officials in a number of cities are empowered by statute to condemn and order demolition of structures found to be a fire hazard. Structures like these are dangerous to the community and are open invitations to arsonists. Condemnation procedures are complex and must be undertaken with expert legal advice.

Notes _____

1. Donal M. Baird, *The Story of Firefighting in Canada* (Erin, Ontario: Boston Mills Press, 1986), p. 18.
2. *Adams v. State of Alaska,* Opinion No. 1318, Alaska Supreme Court, 1976.
3. *State of Alaska v. Jennings,* Opinion No. 1319, Alaska Supreme Court, 1976.
4. *Northern Lights Motel, Inc. v. Sweaney,* Opinion No. 1386, Alaska Supreme Court, 1977.
5. *Wilson v. Nepstad,* Iowa Supreme Court, 1979.
6. *Nolan v. Parker et al.,* Appeals Court of the Commonwealth of Massachuesetts, 1983.

CHAPTER 7

Programs for Public Fire Safety Education

Fire safety education is receiving more recognition than ever before by the fire services of North America. Although public fire education is now a well-established practice in a great many communities in the United States and Canada, it is not yet universal, and more effort must be directed toward this goal.

America Burning, the report of the National Commission on Fire Prevention and Control, recognized the importance of fire safety education. It stated, "Among the many measures that can be taken to reduce fire losses, perhaps none is more important than educating people about fire. Americans must be made aware of the magnitude of fire's toll and its threat to them personally. They must know how to minimize the risk of fire in their daily surroundings. They must know how to cope with fire, quickly and effectively, once it has started."[1]

To get public fire safety education programs under way, there must be a spark plug in the fire department, someone who has attended programs or had experiences that stimulated an interest in fire prevention. This individual can encourage others in the department to recognize the potential of fire prevention. The department's administration must also have a compelling desire to initiate the programs. However it may come about, acknowledgment of the important role of fire safety education among the fire department's responsibilities is a mark of the modern fire service, whether the department be career, volunteer, call, or part paid.

In some cities, fire fighters are assigned the responsibility for conducting public education activities. In Lincoln, Nebraska, for example, fire fighters present fire safety programs to schools, civic groups, and other audiences in their service area. Lincoln's fire department also holds an annual appreciation day for teachers who participate in the "Learn Not to Burn" program and other school fire safety activities. Salt Lake County, Utah, fire fighters and emergency medical personnel present a fire safety and health program in the county's 75 elementary schools.

Scope of Fire Safety Education Programs _____

Fire prevention education is primarily the dissemination of information relating to fire hazards and fire causes in the hope that the public will take proper precautions against fire. To conduct educational programs, fire department personnel should be trained in public speaking and staging demonstrations. In fact, several fire departments have incorporated formal training in public speaking in their recruit training programs.

The term *fire prevention education* generally includes fire reaction related training as well. Chapter 8 is devoted to factors relating to fire reaction; however, a great number of fire prevention education programs include this subject. The term *public fire safety education* includes both activities—fire prevention and fire reaction. Burn awareness education is also included.

Many fire safety education programs have been expanded to include health and personal safety, bicycle and swimming pool safety, drug awareness, cardiopulmonary resuscitation, information on child safety restraint seats, and gun safety. The inclusion of these and other safety subjects must be carefully considered lest the original fire safety message be lost. While all contribute to the welfare of the community, they must be administered in doses that can be easily absorbed by the particular audience.

Public fire safety educators are including wildland fire safety in their presentations in areas where urban-wildland problems exist. The practice of building houses in wooded areas has caused great concern in the fire protection field. Forestry personnel often work closely with municipal fire safety educators to encourage occupants to provide adequate clearances and other safeguards.

Public Fire Education Research

Several excellent fire prevention research projects were carried out with the support of the Public Education Office of the National Fire Prevention and Control Administration, now the U.S. Fire Administration. These projects were designed to shed light on problems in public fire education and to suggest possible directions for improvement.

A preliminary report was submitted by Whitewood Stamps, Inc., as a fire education research study.[2] The study was designed to place fire education, a term encompassing both fire prevention and fire reaction, in context and to delve into high risk as it relates to people, environments, and situations. It also examined the processes by which fire education has been developed.

The report brought out the fact that fire education is a part of the public health and education field as well as the fire protection field. It noted that some excellent fire education programs are carried out by public health and education personnel.

Researchers recognized that "the success of fire education messages depends on the extent to which local conditions are taken into account—e.g., cultural influences, values and beliefs; and types of hazards encountered."[3] They also recognized that the "success of fire education can be limited by the extent of people's control over their economic resources, physical capacities, and the inherent safety of consumer products, as well as by selective perception and competing priorities."[4]

High-risk groups, which often include the young, fire setters, adults, and the elderly were discussed in relation to ignition problems and to death and injury potential. The term *fire setters* refers to juveniles rather than individuals normally thought of as arsonists.

Researchers found that scalds rather than open flames are the greatest burn-related threat to children under three. They found that most young fire setters are boys from homes lacking a strong adult male figure. Supervision and a safe environment were found to be equally necessary to prevent fire-related accidents to children.

Adults were found to be in a high-risk category in relation to ignitions but are in a low-risk category for death and injury because of their competency in handling fire situations.

To the contrary, the elderly tend to be a low-risk group in relation to ignitions but are in the high-risk category in relation to death and injury. A distinction must be made, the researchers believe, between self-sufficient older people and those needing nursing home care and supervision.

Current public education programs were examined and categorized into three types: community action programs, in-school programs, and mass media programs. Five elements appear to be essential to the success of fire education programs: the identification of local fire problems, selection of education strategy, design of materials, implementation, and evaluation of the education program.

Effective use of mass media, it was found, requires knowledge of the value structure of the target population. If messages are to be understood correctly, they must be framed in terms of existing beliefs and attitudes. Localism should be emphasized in the design of educational materials. The importance of having an evaluation procedure was stressed.

This project was followed by the development of a manual entitled *Public Fire Education Planning,* also prepared by Whitewood Stamps, Inc., under the auspices of the then National Fire Prevention and Control Administration.[5] This manual delineates the steps that should be followed in implementing a public fire education program in a community, county, or state.

The manual suggested as a starting point the establishment of responsibility for fire education planning. This includes the assignment of administrative, policy, and staff responsibilities for decisions and activities.

Steps in Public Fire Education Planning

The first step in establishing a program is to identify major local fire problems. These include fire hazards, high-risk locations, high-risk times, high-risk victims, and high-risk behavior. Loss data must be collected as a part of the identification step.

The planner should identify fire hazards by locating records showing causes of fires, determining the most frequent cause of fires, and determining local patterns of fires. Specific local problems can best be addressed by following this planning step.

High-risk locations need to be identified. This includes the determination of which neighborhoods and type of building occupancies in the community have high fire-incident rates. Subsequently, the reasons for these high rates should be determined. Programs should be concentrated in these high-risk locations.

High-risk times should then be identified. Times of day and of year with highest fire-incident rates should be noted. The types of fires occurring at these times should be identified. Fire safety messages should then be concentrated accordingly.

High-risk victims should be identified by the fire education planner. Groups with high fire death and injury rates should be identified. Efforts should be made to determine the reasons for these rates and to involve the high fire rate groups in fire education efforts.

The fire planner next identifies high-risk behavior by determining behavior, either acts or omissions, that causes fires. A decision is then made as to how this behavior can be changed. People are taught proper fire prevention behavior based on these findings.

Baseline loss data must be collected by gathering statistics and information on deaths, injuries, incidents, and property losses due to fire. A determination is then made as to which figures should be used as a baseline for comparisons at the end of the program. Data-monitoring and record-keeping systems must be established.

Major fire problems in the community must be defined. The contributing factors for each serious fire problem must be reviewed. Once this has been completed, the fire educator is ready to move to the selection phase of planning, the second major step.

The fire educator must select the most cost-effective objective for the local program based on available resources. This selection process involves several activities, the first of which is conducting an inventory of community resources. Influential people, media, and civic organizations need to be identified, and personal contact should be made with key people and groups.

An inventory of material resources should be conducted. The availability and cost of fire education material should be determined, and the possible uses of such material explored.

Program cost and benefits must be estimated; alternative programs and estimated costs should be prepared. Possible loss-reduction impacts of each program should be estimated and effective approaches chosen within the limits of local resources.

Educational data must be collected through the use of a questionnaire and a subsequent survey of the community. The questionnaire is designed to determine how much the citizens know about fire safety.

The fire education planner's next step is to select program objectives. There should be agreement on clearly attainable objectives. Objectives should be compatible with local resources, including available personnel. A clear image of program objectives should be presented in order to assist in mobilizing the community.

Effective educational program materials should then be designed and developed. Suitable materials may be available from national fire safety organizations. In the design of material, the first goal is to determine message content. Messages should be directed to specific hazards. They should appeal to positive motives and not be threatening. The messages should show the context of the problem and desired behavior. The format, whether it be a wall chart, film strip, or folder, should be matched to the message, audience, and resources.

The planner should determine the appropriate message time by specifying the target groups and finding out when those audiences will be most receptive to fire safety messages. Messages should be scheduled for maximum effect. *Message,* as used here, includes any type of dissemination of information, whether it be through home inspections, media releases, television, or public fire safety demonstrations.

The program package should be designed and subsequently presented to a sample audience. If found to be effective, the program is then implemented. Material is purchased or produced and distributed, instructors are trained and scheduled, and audience participation and cooperation is obtained.

Programs should be constantly monitored and modified on the basis of the monitoring review. They should be evaluated for effectiveness through review of loss data and educational data. Telephone polls have proven valuable in obtaining such information. As important, however, is the evaluation of program impact. The results should be favorable if the proper steps have been followed in program development.

Home Inspection Program _____

The home inspection program is primarily devoted to fire prevention education. This program probably began in the United States in May 1912 in Cincinnati, Ohio, when the fire department started a comprehensive home and business inspection program. By May 1913 the department had experienced a 60 percent reduction in fires. These inspections were made by personnel detailed from each of the 45 companies in the department. There were 80,000 structures in Cincinnati at the time, and the program was aimed at reaching each of these on an annual basis.[6] Radio did not exist at the time, and fire apparatus was primarily horse-drawn.

Cincinnati has continued an active inspection program through the years. Well over 100,000 building inspections are conducted annually. The majority of these are fire company inspections of dwellings and apartments.

In January 1913 the Cleveland, Ohio, Fire Department started deploying fire personnel from the 35 steamer pumper companies and 13 truck companies to make inspections. During the first nine months of 1913, 249,000 inspections were made, which resulted in the removal of 29,000 loads of rubbish, among other rewards.[7] The Portland, Oregon, Fire Department started home inspections in 1913.

It is interesting to note that Mr. Powell Evans, in a talk to a national fire conference in Philadelphia in October 1913, stated:

> Fire departments, as a rule, all over the country, up to this time have been selected and managed on old "rule of thumb" lines and insufficiently instructed and brought up to modern methods. After emphasizing the need for fire fighting schools, we pass on to our particular subject—the use of active firemen for fire prevention purposes; which provides better occupation for the men, familiarizes them with the property under their control and enables them to perform their hard duty in emergencies with greater speed and safety than otherwise. This system has been employed in a few cities as long as 15 years and is rapidly spreading. It has worked to advantage wherever used. . . .[8]

In 1921 the Wilmington, Delaware, Bureau of Fire started a home inspection program, and Providence, Rhode Island, began in 1930.

During the first week of the program in Providence, which was during fire prevention week, the inspections resulted in the removal of 1,680 tons of combustible material that had been stored in attics and cellars for years. Included were 2,800 discarded mattresses and 1,000 dry Christmas trees. These trees had been stored since the previous Christmas season.

Voluntary Basis

As noted earlier, the home inspection program is primarily a fire prevention education endeavor. Even though the word *inspection* is used, there is no legal backing for the program in most jurisdictions. Entry into residences is on a voluntary basis, and if at any time homeowners look on the program as being mandatory and believe they are being forced to make changes, there is a good chance that it will collapse. Personnel assigned to home inspection duties should be fully apprised of the fact that the program is entirely voluntary.

Training

Fire service personnel must be properly prepared before embarking on home inspection duties. The training program must encompass details of items to be checked, including smoke detectors; the importance of personal neatness and courtesy; and public relations aspects of the job. Successful programs have been carried out in volunteer as well as career and part-paid fire departments.

Training courses should emphasize the approach that must be taken by the fire inspector in an effort to get the message across to the occupants of the homes being inspected. Training as well as evaluation will result in a realization that not all fire service personnel are suited for home inspection duties.

Advance Publicity

To prepare the public to accept home inspections, the program must receive a considerable amount of advance publicity. This publicity should describe the purposes of the program and mention inspection, with particular emphasis on the fact that the program is strictly a voluntary one.

Planning should encompass schedules for areas to be covered each day, and definite routes should be planned to schedule personnel in a way that makes most effective use of their time.

Personnel should be provided with clipboards, flashlights, and inspection forms and be properly attired for contact with the public. There is no reason why nonuniformed personnel, as may be the case with some volunteer departments, cannot participate. They can be provided with identification badges and should wear caps or other identifying clothing for easy recognition by the homeowner.

Senior citizens have carried out home inspection programs in several cities. Under sponsorship of the National Council on Aging, the Fire Department of Huntington, West Virginia, is among those that have used part-time senior fire inspectors. These individuals assist residents by spotting and reporting fire hazards. They have been well received in the community. Residents have recognized that they were there to help and not to enforce codes.

Fire Prevention Education through Civic Organizations

One of the best means of promoting fire safety education is through civic organizations. Organizations such as the Rotary, Lions, Civitan, and Kiwanis are interested in projects that may be of assistance to the community. Most of these organizations have a wide range of community interests represented among their members. This variety of membership is helpful to the fire preventionist who is attempting to solicit the support of the organization. The contacts that can be made through a civic organization can result in garnering of much support for fire prevention measures in the community.

Chambers of Commerce

Another civic organization that can play a part in fire prevention is the chamber of commerce. Local chambers of commerce are usually encouraged to cooperate in fire prevention efforts in the communities they serve. Chambers of commerce may hold meetings in the community to develop greater interest in fire prevention.

The usual method of chamber of commerce involvement in fire prevention is through the organization of a fire prevention committee. This committee should be functioning year-round and should have as its principal goal the furtherance of the cause of fire prevention in the community. The committee should include representatives of the business and industrial community as well as a liaison member from the fire service.

Other Local Groups

Local parent–teacher associations can also be a great assistance in promoting fire prevention programs in the schools as well as in the community. The interests of such organizations are primarily in school fire prevention; however, the groups may wish to have fire prevention education programs as part of their regular meetings. They can be of considerable assistance in supporting the correction of fire safety deficiencies in the schools. Fire department personnel working with the school PTA can also influence the direction of the organization's program.

As a means of promoting fire prevention, some communities have developed programs to appeal to church congregations. These have emphasized fire prevention measures in the church but have also included suggestions that can be taken into the home. Some communities have even staged fire drills in churches in the interest of improving fire evacuation performance. Fire safety messages may be printed in church bulletins.

Fire buff organizations have been instrumental in supporting fire prevention activities in some cities. Members of these clubs represent a cross section of the community and share a common bond of strong support for local fire department activities. Members can be of great assistance in promoting fire prevention activities, in serving coffee for fire prevention open houses, and in raising funds to purchase fire prevention material that might not otherwise be available.

Fire department auxiliaries have been of great assistance in fire prevention efforts in many communities. In some cases the auxiliaries develop and carry out fire prevention programs, while in others they assist fire prevention committees.

Non-English-speaking Groups

In many communities special efforts must be made to ensure that the audience can understand the fire education message being conveyed. Some communities contain large numbers of individuals who are unable to speak or read English. This means that the fire prevention bureau chief must see that programs are designed to reach these individuals. It is helpful to have fire department members who are able to speak the appropriate language. Means must also be available to communicate a fire safety message to those who have difficulty with sight and hearing.

Fire Safety Clinics and Seminars

The fire safety clinic or workshop has been a valuable aid in fire prevention efforts in some communities. Under this concept, representatives of industry and institutions are encouraged to attend a fire safety program during which all phases of fire prevention and reaction are discussed. Some of these programs include outside demonstrations that are most helpful in creating an interest in fire safety.

Seminars of this type are especially valuable if they are aimed at one particular type of audience. If a general audience is anticipated, it is difficult to develop a meaningful program to last for a day. Seminars may be designed primarily for department store personnel, apartment house operators, or nursing home and hospital personnel, as examples.

Statewide fire prevention conferences have been held in a number of states. These are designed to spark statewide interest in fire preven-

Figure 7-1 Fire safety education messages may need to be delivered by signing. (Pam Landis, Partners for Safety, Gettysburg, Pennsylvania.)

tion through the interchange of information on problems and promotion of effective programs. Most states held such conferences in the 1950s as the result of efforts of the International Association of Fire Chiefs. Florida and Maryland have continued to hold annual conferences through the years.

Community Events

Community participation in a fire prevention parade is a good way to alert citizens to fire prevention efforts. Usually the parades are scheduled for national Fire Prevention Week, but in some communities they are held at some other time of the year. Some of the parades consist primarily of fire apparatus; however, fire safety entries should be included.

In several areas, elaborate floats have been developed by fire departments for use in parades. These floats contain messages calling the public's attention to specific measures in fire safety. Generally, they depict a consequence of failing to comply with the message.

Fairs and other organized activities have given a number of communities an opportunity to promote fire prevention. These activities present a chance to set up displays and booths for public informational purposes. They give the fire department an opportunity to have a great deal of personal contact with the general public. Participation at a fair may involve staging displays of fire fighting equipment and fire safety demonstrations. Many county fairs now include excellent fire prevention programs. This gives the department an opportunity to let the public know about its activities.

Inviting citizens to visit a home after a serious fire provides a meaningful safety message. Burbank, California, and several other cities, have held an open house at a recently burned dwelling to let the public see the ravages of fire. This can only be done with the permission of the homeowner; however, many victims welcome the opportunity to help mold positive attitudes toward fire safety.

Fire department personnel must prepare the dwelling by placing signs to indicate point of origin, fire spread characteristics, and other salient features of the fire. Personnel should also be on hand to answer questions.

The staging of the "Burning of the Greens" is another activity that can be most helpful from a fire prevention standpoint. These programs normally include a fire prevention message as well as a message on the potential dangers of discarded Christmas trees that are being destroyed. Air pollution regulations have restricted this measure of promotion; however, burning Christmas trees is still permitted in some jurisdictions.

Learn Not to Burn Program

The "Learn Not to Burn" program of the National Fire Protection Association has been credited with saving many lives. The program entails several segments: public service television announcements, in which brief messages are given by a well-known personality; a "Learn Not to Burn" curriculum designed to be used in preschool through eighth grade; a high school program entitled "Fire Safety for the Rest of Your Life"; Fire Prevention Week, the nationwide fire prevention observance sponsored by the National Fire Protection Association since 1922; "Sparky," the symbol of fire safety for children; and the regional network of fire safety education representatives who assist in implementing the above programs.

The "Learn Not to Burn" curriculum is designed to be taught by teachers in the school system with the assistance of fire service personnel. In some cases, teachers have been unable to adjust their schedules to include "Learn Not to Burn," despite the fact that the local school district desires its inclusion in the overall school curriculum. Volunteer instructors, including fire fighters, may fill this role.

Fire Safety Education in the Schools _____

Opportunities to present fire prevention and reaction messages in schools are limitless. Through the years there have been better chances of getting a fire prevention message to youth groups than to any other segment of the population. The school system should in all cases be consulted before planning any program that is addressed to school personnel or students. The school system, of course, has primary responsibility for educating students, and this responsibility can in no way be abrogated.

A common school fire prevention effort is an appearance before school assemblies. A fire prevention officer can address a school gathering without having to give the talk time and time again in individual classrooms. The appearance before the large group has the disadvantage of limiting personal contact and subsequent questions and discussion. A discussion session can be much more successfully conducted with smaller classroom groupings dealing with age-appropriate material. The opportunity to appear before the entire student body may occur less frequently than opportunities to appear in individual classrooms. The fire department may be able to arrange both types of presentations.

With pre- and post-testing methods, it is possible to assess the relative value of the approaches. There is no question that a consistent, positive program of active learning experiences, reinforced over time, is much more beneficial than a once-a-year "tell them everything you know" presentation.

Although not directly related to fire prevention education for students, a great deal can be said for the inspection of schools as a fire safety measure. The presence of fire department personnel in the school is a desirable feature from a fire protection standpoint.

Home inspection forms have proved to be a highly desirable means of getting across the message of fire safety. Students are given these blanks to take home. They are encouraged to return the blanks after completing them at home. Instructions in completing the form must be given as well as careful feedback on the responses to ensure accuracy. Most students are highly motivated in getting the job done. Parents usually assist with filling in the forms. If such a form is used, thought must be given to the level of understanding and ability of the student. There is no value in using a form that cannot be readily understood by the child filling it in. Home escape plans and smoke detector information may also be sent home with students.

Fire prevention displays developed by students have likewise proven to be helpful in school programs. These displays may be in lobbies or in other frequently occupied areas of the school. To be effective the displays need to be changed fairly often.

Fire Prevention Studies in the Classroom

The fire prevention program for public schools should include training on fire hazards. This training might include information on outdoor fire hazards such as the burning of leaves, the use of cigarettes in forested areas, and indoor fire hazards such as heating appliances, electrical hazards, kitchen equipment, and housekeeping.

The training program should strive to develop a knowledge of fire safety measures that might be employed by young people. Fire prevention personnel have sometimes misdirected training efforts by not considering the learning and comprehension abilities of the students being taught. There may be a problem whereby an interest in trying out fire is sparked in certain age groups. Psychologists have indicated that many people have a natural curiosity about fire, especially younger children. Fire prevention efforts should not encourage this interest.

It is possible at the junior high or middle school level to include fire prevention in other courses. History classes, for example, could cover some of our major fires to develop a greater realization of fire as a problem in the country. Likewise, civics classes might include discussion of the fire department as a part of the city government, with emphasis on the cost of fire protection in the community as a part of the municipal budget. Factors relating to the complete fire picture should also be included. The impact of fire insurance might be mentioned as well. Likewise, fire safety might be included in the study of economics. The economic impact of fire on society could well be a subject of interest in an economics class.

Fire prevention may be taught in science classes. Experiments in the chemistry of fire could be of considerable help in teaching fire safety at the middle school and junior and senior high school levels.

Certain fire prevention subjects could be included in home economics classes. Students might be taught about the use of flame-retardant textiles and about the need for fire evacuation plans in the home.

Shop subjects can also include fire prevention. There are many opportunities for the inclusion of fire prevention in shop subjects because they often include the use of flammable liquids as well as the storage of rags and other combustible materials.

The overall program for fire education in the school system should include some contact with the subject at each grade level. Unfortunately, personnel limitations and other factors often limit the activities to one or two grades.

Some states have had mandatory fire education programs for a number of years. Ohio passed a law in 1908 requiring "that every teacher in every public, private or parochial school shall devote not less than thirty minutes in each month during which the school is in session, to instruction of pupils between the age of six and fourteen years, in fire dangers." The state fire marshal was required to prepare a textbook to be published by the state and provided to each teacher.[9]

At least 35 states require fire safety education in the schools. Implementation of required programs is not well enforced in some states, however, and many feel that locally developed nonmandatory programs are more effective.

Coordinating for Success

Close coordination with school faculty is necessary for the development of successful fire education programs in the schools. Many school systems have a standing policy against including material in the curriculum that has been developed outside the system. A spirit of cooperation with appropriate personnel in the school system may result in modifying such a position to provide at least some fire prevention coverage where no material has been developed in the education system.

Another problem in school fire prevention programs has been with the distribution of commercial fire safety material in the schools. Fire departments in a number of communities have asked the school system to permit the distribution of fire prevention literature published by insurance companies, bottling works, candy companies, or other commercial enterprises. Although because of budgetary limitations the fire department may have no choice but to use this material, the school authorities in some communities do not feel that they can permit the distribution of fire safety material advertising an individual commercial organization.

Future teachers in the elementary education program at Chadron State College in Nebraska receive a four-hour presentation at the local volunteer fire station each semester. This program, which has been in effect since 1984, introduces the students to several fire safety education programs, including "Learn Not to Burn." The program is designed to encourage the future teachers to teach fire prevention and safety education once they reach the classroom. A deputy state fire marshal teaches the classes.

Junior Fire Marshal Program

The Junior Fire Marshal program has been effective in many sections of the United States. Both English and Spanish versions are available. In some areas this program has been sponsored by a local insurance organization, with the field work done by fire department personnel. In most localities the material has been supplied by the school district or fire service, and all field operations have likewise been under that agency. Membership in the Junior Fire Marshal program is usually obtained by filling in a home fire prevention form or by taking some other step that aids the cause of fire prevention. School students enrolled in the program are usually given certificates of fire safety knowledge.

Sponsored by the ITT Hartford Group, the Junior Fire Marshal program is designed for children in kindergarten through third grade. It focuses on fire prevention tasks through the use of teacher guides, films, filmstrips, and audiocassettes. Fire safety in apartments as well as single-family homes is addressed.

Hospital Programs

Many fire departments have become involved in the National Safe Kids Campaign. This organization, through state and local chapters, promotes bike helmets and bike safety, child vehicle occupant protection, and general injury prevention as well as burn prevention and fire safety. The campaign is a program of the Children's National Medical Center in Washington, D.C.

Burn prevention programs operated by hospital burn units may be successfully combined with fire department prevention programs. This combination is used successfully in school programs staged by hospital burn units and the local fire departments. Burn awareness activities of the American Burn Association may be implemented by fire services as well.

Scout Groups

A number of fire departments have had considerable success working with Scout organizations. This work includes helping Scouts earn merit badges, making appearances at Scout meetings, giving demonstrations before Scout gatherings, and arranging visits to fire stations by Scout groups. In addition Scouts may participate in wastepaper drives, salvage drives, and other efforts that result in improved fire prevention practices. Another worthwhile procedure has been the use of Scouts for fire hazard hunts in which Scouts go into homes in the community to look for hazards. A prize is awarded to the group making the best inspection.

Some Scout groups have assisted in beautification programs. They plant trees, clean up parks, and remove hazardous vegetation. Scout groups have also assisted with baby-sitter training programs. The baby-sitter training programs normally given by fire departments can be readily adapted to Scout groups. Programs can also be adapted to 4-H groups and other youth groups. Such programs may include specific information for farm fire safety problems.

Public Safety Combined Programs _____

Public interest in crime prevention is generally greater than that in fire prevention. For this reason, some jurisdictions have combined crime prevention and fire prevention education programs. Cities with public safety departments have found this to be a desirable concept. In addition, some communities recognize that the police department is often more welcome delivering a crime prevention message and therefore have police and fire personnel participate in the same session. Usually, the program starts with crime prevention, and then the fire service person concludes with a fire safety message.

An example of a program in which fire safety and other public safety issues are addressed is conducted in North Dakota:

> The Fire Marshal's Office joined forces with two other state agencies that needed to get information to the state's senior citizens, the Bureau of Criminal Investigation (BCI) and the Consumer Fraud Division. BCI had a need to address crime affecting the elderly and Consumer Fraud had a need to make elderly consumers aware of the great number of "rip-offs" involving senior citizens. The three state agencies put together a program called "The Protection Connection" and made presentations across the state. The vast majority of presentations were made at senior citizen centers. This approach seemed to work very well.
>
> The fire safety portion of the program was divided into two segments. The first segment dealt with fire prevention and included the topics of: electrical safety, kitchen fire safety, fire safety in the living room and bedroom, and general fire safety. The second portion of the program focused on surviving a fire and steps to take when a fire occurs. Statistics and fire facts were used to emphasize the seriousness of the problem. The next portion of the program dealt with smoke detectors including testing, changing batteries, and proper placement. The next topic was the need to develop a fire escape plan. Emphasis was placed on getting out fast when a fire occurs, having two ways out, having a meeting place outside the home, and a plan to call the fire department. "Stop, Drop, and Roll" was also a part of the presentation. At the end of the program, time was allowed for questions and answers. On many occasions this segment turned into a kind of "show and tell" with the audience sharing experiences they had with fire. We concluded the program by distributing several hand-outs including "Fire Safety Tips for Older Adults," published by NFPA.[10]

Publicity Programs _____

A variety of programs can be devised to capture the attention of the community and to publicize the theme of fire prevention. Following are

Figure 7-2 Fire safety activity centers feature interactive, hands-on displays teaching fire prevention behaviors. (Fire Museum of Texas, Beaumont.)

some ideas that have been used by fire departments in the United States and Canada.

Contests

Poster contests have been widely used in the fire prevention field for many years. These contests are generally conducted by the local fire department or fire service association in cooperation with community schools. The contest should develop an interest in fire prevention and not alienate school officials from the fire service. Some school officials believe that poster contests are inappropriate as a means of achieving fire prevention because unfavorable educational relationships may be created by such contests. The artwork in these contests may not be in keeping with educational programs. Any fire department considering a poster contest should first check with school system officials and obtain their sanction before proceeding. Entries must be checked for technical accuracy.

Promotional Aids and Activities

Place mats bearing a fire prevention message have also proven to be effective. Diners will often read the wording on a place mat while waiting to be served. The use of table tents with printed messages can also be helpful. The table tent normally does not have to be destroyed after each meal as does the place mat.

Favors such as balloons, rulers, and fans can be valuable in promoting fire safety. These favors are usually relatively inexpensive and are appreciated by children as well as adults. They are often passed out at public events as prizes for answering fire prevention questions accurately. Favors should not present a safety hazard to recipients.

Some communities have used a loudspeaker on a fire truck to get out a fire prevention message. The fire engine is driven around the community to attract interest. Citizens on the street are given fire prevention messages.

The hydrant collar is another fire prevention message-bearer used in some communities. Talking hydrants, which are operated by remote control, have proven to be very effective in imparting fire safety messages to young children.

Stenciled messages on public streets have been quite helpful as well. These stencils spell out appropriate fire prevention messages and are changed periodically. Placards on safety islands, lampposts, and other city property can likewise be helpful in fire prevention efforts. Posters are also used on fire department vehicles, municipal buses, and municipal garbage trucks. Cities with subway systems have a great opportunity for presenting fire prevention messages through the use of placards. These can be seen by thousands of people who travel on the subway each day.

The use of pay envelope enclosures has proven to be effective in some areas. Milk cartons contain a panel that can be used for rather complete fire prevention messages. Bread wrappers can also be printed with fire prevention messages. Some communities, with the cooperation of local dry cleaners, have taken the message of fire prevention to the public on dry cleaning garment bags.

Outdoor reader-board and billboard fire prevention publicity can be helpful. At least one major grocery chain also uses these boards for fire prevention messages.

Static fire prevention displays have proven successful in some communities. These displays may be placed in store or bank windows or in other locations where they can be readily observed by the public.

A number of cities have opened permanent facilities in which citizens can learn fire prevention measures as well as fire survival procedures. Milwaukee, Wisconsin's Survive Alive House, a converted recreational field house, provides visitors with an opportunity to practice home fire drills under realistic conditions. In Rome, Georgia, a fire department substation was constructed with a fire safety education center as an integral component.

One of the most successful citywide fire safety activities is the Great Louisville (Kentucky) Fire Drill, which has been held annually on Sunday of Fire Prevention Week since 1984. The drill is designed to educate the public in fire safety by giving participants a chance to practice home fire drills and learn to stop, drop, and roll. They also learn about matches and lighter safety and how to report a fire. From 12,000 to 15,000 people attend the event each year. Many other exciting activities take place at the Drill, which rotates among the city's parks so that all neighborhoods have access to the event.

Fire clowns are quite successful in teaching young children basic fire safety such as "Stop, drop, and roll" and home evacuations. Florida's state fire college conducts weeklong classes to meet the need in this specialized field.

Media Publicity

An important part of any fire prevention undertaking is obtaining publicity through the media. Press releases must be interesting, direct, and to the point. Material that is uninteresting and vague will not receive newspaper, television, or radio publicity. It is necessary for the public information officer, chief, fire marshal, or other individual connected with fire safety publicity to work to develop or locate material that will be accepted and used by the media. The fire prevention officer should visit the media offices to develop proper rapport with their staff. Such a visit may be quite rewarding and can result in long-term cooperation between the media and the fire department.

Any fire safety proclamation that might be developed in cooperation with the mayor's or governor's office should be included in press coverage by the fire department. This includes fire prevention and cleanup week proclamations.

The press should also be encouraged to accurately report fires, and particular attention should be given to including information on contributing factors. There is no greater opportunity for disseminating fire safety information than by reporting fires in the community.

Any special events carried out by the fire department should be reported to the media. Many times the department is discouraged when articles do not appear immediately; however, the news media can only use material of a nonurgent nature on occasions when their newsrooms are not already filled with information having a higher priority.

The Denton, Texas, Fire Department prepares a special supplement to the local newspaper that is published each year during Fire

Prevention Week. The supplement includes information on the department as well as a considerable amount of fire safety advice.

Radio stations can contribute a great deal to fire prevention publicity. They can run public service spot announcements, which have been found to be among the most effective means of getting out fire prevention messages. Announcements may be developed by the fire prevention bureau locally or by one of the national fire organizations. Locally prepared announcements are usually the most effective; however, some announcements prepared by national organizations are designed to permit wording pertaining to local problems. Messages may address current or seasonal fire problems.

Television stations, likewise, can include announcements and other publicity that can help the cause of fire prevention. Spot announcements can be very effective, even though they may be quite short. The fire department can obtain excellent publicity by maintaining close liaison with the radio and television stations in the community.

In many communities, the fire department has been successful in getting television coverage of fire safety demonstrations and other such activities. The fire department will probably not always have prime time available for such promotional programs.

Radio interview programs can be of considerable value when they use fire prevention personnel. Interviews can be used to highlight particular problems faced by the community.

On occasion, motion-picture theaters have cooperated in fire safety education programs by placing messages on marquees, showing short fire prevention films, and permitting the staging of live fire prevention demonstrations in the theater.

An effective fire safety publicity measure is a bus or trailer. Such trailers or buses can be equipped to include a considerable amount of material for demonstration and may be arranged to allow rapid movement of people through the display. Many communities have outfitted such a unit. In some cases the unit is a surplus bus from the municipal transit company. The limited mileage expected of the unit allows vehicles to be used that may no longer be fully suitable for first-line transportation service. A talking Sparky can be added to the fire prevention bus or trailer to attract attention to the unit. A number of the trailers are equipped to demonstrate smoke detector and residential sprinkler operation.

Smoke Detector Programs

There are many smoke detector distribution programs. In some instances, smoke detectors are installed in the home by fire fighters. This program is very easily incorporated into a home inspection program; however, the liability issue should be addressed before program implementation.

In some areas, smoke detectors are distributed to groups and locations that are considered to be high risk. For example, the National Safe Kids Coalition is involved in distributing smoke detectors to homes where young children reside. Young children are at risk because they are not capable of self-preservation in the event of fire. Some senior citizens and the disabled are also considered to be high risk because of impairments that make it difficult or impossible for them to react appropriately in the event of fire. Low-income residents are provided with detectors in many communities.

Because smoke detectors provide early warning when fire occurs, this type of program can impact fire loss data in a community. Any smoke detector program should be ongoing and should address smoke detector maintenance and battery replacement.

Fire Prevention Week

Fire Prevention Week has long been looked on as an opportunity to spotlight fire prevention for all citizens during a concentrated time span. The observance began as Fire Prevention Day through a proclamation by President Woodrow Wilson for October 9, 1920. In 1922 the observance was extended to one week by proclamation of President Warren G. Harding. October 9 was chosen to commemorate the great Chicago fire in 1871.[11]

Fire departments throughout the United States and Canada use Fire Prevention Week as the time for aggressive activities aimed at all segments of the population. Radio, television, and newspaper announcements are extensively used at this time. Fire Prevention Week was for many years the high point for the year's fire prevention activities in a community.

Volunteer Fire Departments

Rural volunteer fire departments can also have excellent public fire safety programs. In southeastern Oklahoma, a number of rural fire departments join with private businesses, the State Forestry Service, and the U.S. Forest Service to provide fire safety education in their communities. These programs address arson prevention in the woodlands as well as structural fire safety.

While it is difficult for many volunteer fire departments to conduct mandatory fire safety inspections, no conflict of interest arises with public education activities. Volunteer fire services usually have individuals who can best use their talents in public education work rather than fire suppression work. This is true of rural, urban, and suburban volunteer fire departments. In fact, one of the outstanding public fire safety education programs is that of the Wayne Township fire department, a volunteer organization serving a large suburban area near Indianapolis. The department's volunteer public education staff is complemented by volunteer fire investigation and code enforcement divisions.

Review of Successful Programs _____

Proving Public Fire Education Works, a publication of Tri Data Corporation, describes 77 public education programs in the United States and Canada. Comprehensive communitywide and school programs, specific causes and target groups, juvenile fire setter and smoke detector programs, as well as national programs and special topics such as those related to wildfire prevention are covered. Practically all resulted in a reduction in fires, injuries, and fatalities.

The publication includes a summary of factors in common that make programs successful:

- They have "champions" who see the program through and lead its implementation.
- They are situated in departments with magnanimous chiefs who allow their public educators room to be innovative and to seek outside resources.
- They carefully target a particular aspect of fire safety, or strike in force across a broad front over and over, reaching a large percentage of the population.
- Market research in one form or another is used to tailor the programs to their intended audience. Powerful allies are obtained, often in the business community or the education community, to get through the bureaucratic barriers and provide assistance.
- The materials used in the program may not be fancy, but they are clear and in abundant quantity.
- The programs reach a significant percentage of their target audience with public educators often going door to door, literally, or through the media to have a broad impact.
- They often repeat messages over and over just as an ad campaign would.
- The good programs are adaptable, changing goals and materials as the fire problem changes.
- And they often are refined by testing in a small area for a small target population before they are implemented communitywide.[12]

A survey to determine attitudes toward and knowledge of fire safety by the public and fire fighters can be quite helpful. Steve Bethke, Public Educator for the Minneapolis, Minnesota, Fire Department, conducted such a survey. In his canvass of neighborhoods, he found that the average citizen felt that 24 percent of the department's budget is spent on fire prevention when the actual amount is 6 percent. Another question asked whether the respondents felt the fire department was likely to arrive in time to save anyone trapped in a burning building. Eighty-seven percent felt that this was a true statement, indicating a need for greater public education emphasis on this issue as conditions seldom enable such rescues. The Minneapolis survey also found that the fire fighters surveyed strongly support the need for public fire education.[13]

The fire department in Des Moines, Iowa, used a similar type survey to help focus on fire safety education goals. This survey, directed at school students, indicated that only 54 percent were aware of whether portable heating appliances in their homes had safety switches to shut them off if knocked over. The survey indicated that 78 percent had working smoke detectors in their residences.

Summary _____

The field of public fire safety education offers many opportunities to reduce losses of life and property by fire. The key to success in the field is not always easy to find. Hard work and enthusiastic support are necessary ingredients.

Fire safety education is the dissemination of information relating to fire hazards and fire causes with the hope that the public will get the message and take the proper precautions against fire. It also includes the development of proper fire reaction. Among the avenues used to reach the public is the home inspection program, which apparently had its beginnings in Cincinnati, Ohio, in May 1912. Although the program is called "inspection," participation is voluntary on the part of the homeowner.

The fire prevention educator can also work through civic organizations, such as chambers of commerce, parent–teacher associations, clubs, and churches, by giving talks and demonstrations. Community events—parades, fairs, dances, beauty contests, and other organized activities—offer many opportunities to publicize the message of fire prevention and enlist community cooperation.

Fire safety education in the school classroom offers limitless opportunity to instill lifetime attitudes about fire prevention precautions. Students who are taught to be aware of hazards that cause fires and what to do about them can help the cause of fire safety in the home and in the community. School programs can take the form of actual classroom instruction or talks and demonstrations before school assemblies. The fire preventionist can also work with young people in group activities outside of school, such as junior fire marshal programs, Sparky fire departments, and the Scouts.

Publicity aids for fire prevention can take many forms: contests staged by the fire department; billboard advertising; distribution of signs and posters, book covers, favors printed with fire prevention messages; innovative ideas like talking trash cans or fire alarm boxes; or roving fire engines delivering messages through loudspeakers. The possibilities are endless.

One of the most effective means of reaching the public is through media publicity—newspaper, radio, and television. The press should be encouraged to report fires accurately and to give specific information on contributing causes. This can have a tremendous impact on citizens of the community. Public service spot announcements on radio and television and television coverage of fire safety demonstrations and special events can all play an important part in making the community aware that it is everyone's duty to help in preventing fire.

Notes

1. U.S. National Commission on Fire Prevention and Control Report, *America Burning* (Washington, D.C.: U.S. Government Printing Office, 1973), p. 105.
2. Whitewood Stamps, Inc., *Fire Education: People, Programs, and Sources,* prepared for the National Fire Prevention and Control Administration (Newton, Mass., 1975).
3. Ibid., p. 3.
4. Op. cit.
5. Public Education Office, National Fire Prevention and Control Administration, *Public Fire Education Planning, a Five Step Process,* prepared by Whitewood Stamps, Inc. (Newton, Mass.).
6. Powell Evans, comp., *Official Record of the First American National Fire Prevention Convention* (Philadelphia: Merchant and Evans Co., 1914), p. 150.
7. Ibid., p. 384.
8. Ibid., p. 150.
9. Percy Bugbee, *Men Against Fire: The Story of the National Fire Protection Association, 1896–1971* (Boston: National Fire Protection Association, 1971), p. 20.
10. Personal communication to the author from Robert Allan, North Dakota State Fire Marshal.

11. Bugbee, p. 39.
12. Philip Schaenman, Hollis Stambaugh, Christina Rossomando, Charles Jennings, Carolyn Perroni, *Proving Public Fire Education Works* (Arlington, Va.: Tri Data Corp., 1990), pp. 113–115.
13. Steve Bethke, "Public Perceptions in the Marketing Mix," unpublished paper, 1993.

CHAPTER 8

Instilling Positive Fire Reaction

The reaction of the individual in a fire emergency is a critical element in life safety considerations, and the human factor should be carefully analyzed in the development of fire safety codes. Public fire reaction is an integral part of fire prevention planning and education.

Any assessment of anticipated human reactions in the event of a fire must take into account individuals of all ages, health categories, physiques, occupations, and temperaments. The reactions of individuals to fires vary in relationship to these characteristics and to other factors such as the severity and location of the fire, the manner in which it started, the presence of other individuals within the fire area, and the specific condition of the individual at the time. The latter consideration includes factors such as whether the individual is awake or asleep, is intoxicated or under the influence of drugs, or is fully alert and capable. It also includes any other temporary conditions that may have an effect on the individual's reaction to the specific conditions with which he or she must cope.

A number of programs carried out in the field of fire safety by fire departments, as well as those carried out by other organizations, are in fact reaction-type activities.

Individual Decisions when Fire Occurs _____

Bernard M. Levin, Ph.D., formerly of the National Institute of Standards and Technology, has studied fire reaction for many years. He has concluded that people can usually be expected to make rational decisions during a fire; however, they may make inappropriate decisions that increase their danger because they do not understand the nature of fire and its growth and the elementary principles of fire protection.

Levin suggests that fire safety education programs address the phenomena of flashover; the speed with which a fire that seems small can grow to flashover; the concept of margin of safety; the size and nature of fires that building occupants can extinguish; and the value of closed doors and danger of opening doors. He explains the concept of "margin of safety" as the time between when everyone has left the area and when anyone in the room would be injured or killed.[1]

Fire Reporting Procedures

An example is the ongoing campaign to instill in the public an ability to properly summon the fire department. In addition to using stickers on telephones and signs on fire alarm boxes, some cities have resorted to billboard advertising and placing emergency telephone numbers on the sides of fire department vehicles in an effort to increase public awareness of fire reporting procedures. An effort is made to train individuals so that they will immediately report a fire. The time factor is emphasized in these campaigns, in an effort to encourage individuals to render a report without delay. An effort is also made to ensure that the proper number is used if the fire is reported by telephone and to ensure that people give the correct address and other necessary information to the dispatcher. The adoption of the enhanced 911 reporting system has been very helpful.

Home Fire Drills

Another example of fire reaction efforts is the home fire drill program. This outstanding program, which has been under way for a number of years, was given considerable impetus by the Ohio State Fire Marshal's Office in the late 1950s. At the same time, a major effort was undertaken by the Kansas City, Missouri, Fire Department, which resulted in widespread adoption in the greater Kansas City area. This program has national stature and is being carried out in hundreds of communities as part of the activities of the annual Fire Prevention Week in October, which is proclaimed by the president and endorsed by the National Fire Protection Association and other national fire safety associations. Members of the public are encouraged to conduct a home fire drill at a predetermined time during Fire Prevention Week. In many communities the program is known by the acronym EDITH: "Exit Drills in the Home."

The program encourages individuals to locate potential escape routes and to make plans to leave the house without delay in the event of an emergency. This means determining all possible avenues of emergency egress and training family members to leave by these doors or windows. The program also emphasizes the importance of leaving doors to occupied bedrooms closed to eliminate smoke travel as much as possible, although widespread use of corridor smoke detectors has raised a question as to the validity of this practice. Under the home fire drill program, individuals are taught to feel doors at their highest point before attempting to open them in the event of fire. They are taught that the highest point of the door is most likely to be the hottest if there is a fire on the other side. Participants are advised to check for smoke, and if it appears safe, to enter the corridor crawling on hands and knees, if necessary, to stay in the portion of the corridor that has the least smoke buildup.

Another important part of the home fire drill program is a consideration of the need for all family members to assemble at a predetermined point once they are outside. This assembly gives the fire department personnel who arrive first on the scene a means of learning which rooms might still be occupied to give them a better chance in any rescue operations. Fire department officers can immediately determine whether rescue operations are necessary. The officers will be able to check with a family member to determine where victims might be located.

The home fire drill program emphasizes training in turning in fire alarms and the need to turn in an alarm without delay, although it recognizes that there are times when it might be more appropriate for the individual who discovers the fire to notify occupants before turning in the alarm.

Home fire drill training is a logical sequel to smoke detector publicity. The tremendous popularity of smoke detectors has resulted in the production of hundreds of pamphlets, instruction sheets, and guides relating to this subject. Many of these have additional coverage on home fire drills. The reader is encouraged to practice a home fire drill and plan immediate evacuation if a fire is detected by the smoke detector.

Closely related to both programs is training in procedures to follow when clothing catches on fire. People are taught the simple procedures of rolling over to control the flames. The procedure is known as "Stop, Drop, and Roll." These steps have saved many lives and minimized injuries in countless incidents.

Although these have proven to be extremely successful programs, countless communities for one reason or another have not used them. Incidents where home fire drills have paid off have been publicized in local communities; however, many incidents have undoubtedly gone unnoticed in which home fire drill procedures have proved to be effective and, in fact, have resulted in the saving of life.

The baby-sitter training program is an adjunct of home fire drill efforts to instill proper public fire reaction. This program includes other facets of safety; however, the fire prevention and fire reaction

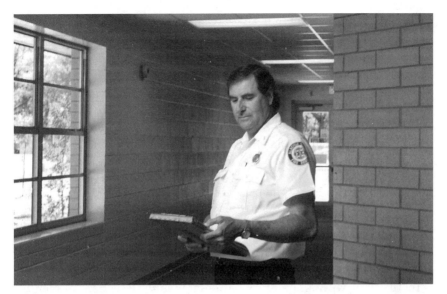

Figure 8-1 School fire drills have more impact when witnessed by fire service personnel. (Alachua County, Florida Fire-Rescue Department.)

function is one of major importance. Training is given in measures to reduce the possibility of fire as well as in procedures to follow should a fire occur. Baby-sitters are instructed in means of evacuating infants and small children, in fire reporting procedures, and in other measures relating to fire reaction.

School Fire Drills

Probably the most widespread program to condition public fire reaction is the fire drill program carried out in schools throughout the United States and Canada. Practically every state and province has requirements for fire drills to be held on a periodic basis in public, private, and parochial schools. These drills are designed to ensure, so far as possible, the immediate evacuation of students from a school building in the event of a fire or other emergency. The schools accept the fire drill program as a part of their operation and have shown little hesitation in actively requiring drills.

Many hidden benefits have come from the school fire drill program. Undoubtedly, individuals have reacted properly to fires in places other than schools as a result of training they received while they were in school. Fire drills in schools have probably helped individuals more readily to accept fire drill programs in institutions, industries, and other locations. Reactions to fires in theaters and other public assembly locations are also more positive as a result of conditioning received in school fire drill programs.

Institutional Fire Drills

Fire drills in institutions, although probably not as widely publicized as the school drill program, have many similarities, as well as certain differences. In the institution, especially where nonambulatory individuals reside, reaction is usually expected only on the part of the institutional staff and visitors. The patients most often are not counted on for reaction of any kind, and institutional personnel initiate all measures. Complete evacuation, which is normally achieved in educational occupancy drill programs, is not generally expected in institutions. The drill may include horizontal evacuation to a place of safety within the structure rather than to a location outside the building.

Institutions with ambulatory patients, including sections of mental, drug, and alcoholic institutions, may have programs in which each individual is part of the fire drill. Patients' participation in the fire drill will depend on their individual abilities.

Typical evacuation procedures are obviously not practiced in places of incarceration. In such institutions, measures are taken to eliminate as far as possible the chance of escape. This means that evacuation is at best most difficult and must be planned on an individual basis in each facility. In places of incarceration, assigned personnel must assist individuals in evacuating the building. Although inmates in a prison are most likely physically capable of departing from a burning structure, they must first be released from their cells by correctional officers.

Industrial Fire Drills

Many years ago little emphasis was given to fire drill programs in industry. Long hours coupled with arduous tasks left little time for fire drills. The 1911 Triangle Shirtwaist fire, as well as several other fires, including the Hamlet, North Carolina, fire on September 3, 1991, have indicated the need for comprehensive fire reaction programs in this class of occupancy.

In arranging fire reaction programs for industry, officials must remember that shutdown time is expensive and that many machines and processes cannot safely operate unattended. Personnel can be familiarized with the location of fire exits, the location of fire alarm stations, and procedures to follow in the event of fire. As in the case of institutions, employees in industry may have duties to perform in reaction to fire that involve more than merely evacuating the structure. They may be responsible for obtaining and operating fire extinguishing equipment and calling the fire department.

Proper Use of Fire Extinguishers

Probably a major weakness in fire reaction programs is failing to adequately train people in the use of fire extinguishers and in the limitations in their capabilities. Records are replete with incidents of individuals attempting to control fires with extinguishers that were inadequate or improper for the conditions encountered. In cases where the individual also delays in reporting the fire to the fire department while using the extinguisher, fires resulting in large losses may ensue.

A great deal of thought has been given to the relative role of the portable fire extinguisher in the fire protection field. This device depends entirely on the proper human fire reaction for any effectiveness. Directions printed on the fire extinguisher are sometimes detailed and may be somewhat intimidating to an individual who has to use it.

This factor also influences the effective use of fire extinguishers as fire suppression devices. Programs for education in the use of fire extinguishers are widely available. Only through such training could all of the portable fire extinguishers in buildings throughout North America be effectively used by the first person to spot a fire and an adjacent fire extinguisher. The all-class fire extinguisher, which may be used on class A, B, or C fires, has simplified training procedures considerably.

Traditionally, training in the use of fire extinguishers has been confined to individuals who have a direct responsibility in protecting specific locations, such as nurses in hospitals, teachers in schools, or maintenance employees in industrial plants. Even in industry, training of this type is not normally given to individuals who have little likelihood of using an extinguisher. Likewise, office workers in industry seldom are given such training.

Several communities have moved to eliminate fire extinguishers from certain public places. This action was inspired by the problems connected with training people and in recognition of limitations of fire extinguishers. The best solution would be to conduct training programs for occupants of these structures. Fire extinguishers have extinguished thousands of fires in the incipient stage.

Large Residential Occupancies _____

Another place where fire reaction training could enhance fire safety is the multifamily residential occupancy. Apartment houses and condominiums with a number of units generally have interior fire alarm systems as well as fire extinguishers mounted at various locations in the structure. There are very few apartment houses or condominiums in which drills are carried out or in which fire extinguisher training is given.

Individuals living in apartments generally give little thought to the fire safety features of the structure. If a fire drill were staged periodically within the building, tenants might feel a sense of security. Other types of residential operators may be more inclined to hold fire drills on a periodic basis. For example, fire drills are fairly common in college dormitories. This procedure has saved lives on many occasions. Training of employees in fire alarm activation and fire extinguisher use is very important in the hotel industry.

Prevention and Control of Panic Reactions _____

A major consideration in educating the public for proper response is that of potential problems created by panic. In countless incidents panic has played a major part in causing a large loss of life. Incidents include real and supposed fires as well as emergencies of other kinds that have occurred through the years.

In planning for public fire reaction, the planner must consider conditions that cause groups to panic, as well as the steps that may be taken to reduce the effects of panic.

Conditions that Contribute to Panic _____

Conditions that cause a group to panic include the existence of a real or imagined danger. Individuals in the midst of a panic usually are taken by surprise by the condition with which they are faced and may make a hasty evaluation of the possible dangers. They may see flame as a sudden "flash" creating an impression that a major fire exists, while in fact there is actually little danger because the flash may be extinguished momentarily.

Of course, in many cases there is a real danger to the individual; however, reaction to the danger in a panic may make the difference between life and death. For example, a fire might occur in a building with combustible ceiling tile. If the occupants are able to evacuate the building in an orderly fashion without delay, they can well escape the dangers of the burning tile. On the other hand, sudden movement toward the exit could bring death if this panic were to cause individuals to fall. Those people who have fallen could preclude other occupants from safely leaving the structure.

Another consideration in a panic is personal condition, such as malnutrition, monotony, intoxication, fatigue, and other factors that affect the physical well-being of the individual. An individual in a state of great fatigue or suffering from malnutrition is going to react differently in a panic than is an individual who has no such problem. Psychological factors must also be considered in determining the possibility of panic occurring. People with feelings of insecurity, or an inclination to worry a great deal, or in a general state of tension may act unpredictably in a panic. These factors make it difficult to assess conditions that might create panic.

Measures to Counteract Panic

A number of positive steps can be taken to reduce the possibility of panic. The importance of a few seconds in a panic cannot be overemphasized. On several occasions an individual has taken control of a panic by immediate action. For example, in a theater fire some years ago in Florence, South Carolina, a young usher went to the front of the theater, announced a fire drill, and asked everyone to leave in an orderly manner. Many people thought that this was in fact only a drill. They reacted calmly and departed the structure only to realize that there was a major fire in the building. Quick thinking on the part of an individual can be quite important, especially if the individual can be identified as having some authority. For this reason certain fire departments always have uniformed personnel on duty in theaters in their cities.

Uniform as Symbol of Authority. The uniform has been found to be of considerable value in bringing about positive reaction to potential panic situations. In an explosion in Perth Amboy, New Jersey, some years ago, a uniformed milkman was looked to for advice and guidance by many people because his uniform represented authority and proper direction. Uniformed fire service personnel have been helpful in several other similar incidents.

Posting of Maximum Occupancy Signs. Posting structures for maximum occupancy may also help prevent panic. Posting should be based on measurable standards so that the number of people inside the structure at no time exceeds the exit capacities and predetermined limitations based on a square-footage formula.

Overcrowding is a very difficult condition to control, especially where people are moving in and out randomly. This type movement makes it extremely difficult to properly record the number of people who happen to be in the place at a given moment. Of course, panic possibilities are not limited to places that are enclosed or under cover. Panic may occur even in an outside location, and every consideration must be given to providing adequate means of egress and emergency notification and to preventing overcrowding.

It may be difficult for management to enforce fire capacity requirements in places where popular performances are being held. A fire official becomes unpopular in attempting to enforce requirements of this kind. This is especially true where the audience is excited as a result of the program being held. Failure of the feature attraction to arrive also creates problems. It may be necessary to prohibit additional patrons from entering the facility. The facility manager should determine who must leave the premises if overcrowding must be relieved.

Emergency Lighting. Emergency lighting has also been found to be helpful in preventing panic. Many jurisdictions are requiring constant charge battery-operated emergency lighting units designed to give at least a minimum amount of light during a power failure. Exit lights are also important for preventing panic. Practically all building and fire codes require exit lighting throughout by use of a lighting system designed specifically for marking exits. Under electrical code provisions, these lights are wired in such a manner as to ensure continued operation except in the event of a major power failure. Exit lights, from a circuit that comes off ahead of the main disconnect for the building, provide continued operation even if the building electrical system experiences internal failure.

Exit Signs. Much study has been given to proper wording and color for exit signs. Some codes require green, while others require red. The prevention of panic has been a factor in these deliberations and in the development of requirements. Some people associate red with fire and with exit lights. The other school of thought is that green indicates safety and that an individual seeing green will automatically go toward that door to reach a place of safety. There are valid arguments on both sides. For that reason the Life Safety Code of the National Fire Protection Association (NFPA 101)[2] recognizes both colors.

A combination exit light and emergency light is on the market. This device provides directed lighting under the fixture for exit visibility and, in addition, provides lighting of the exit sign.

Maintenance of Exit Facilities. An important measure in panic prevention is maintenance of exits. No exit can be considered usable if it is necessary to obtain a key or remove obstructions. Individuals will react differently under the stress of even a small amount of smoke or heat. While it may be a minor problem to have to remove a chair that is blocking an exit, in time of stress this may be a monumental task. A chair or other blockage may cause a number of deaths because people cannot be expected to act rationally under such conditions.

Another example is the mistaken idea of some public assembly operators that certain exits may be locked when only a less than capacity audience is anticipated. Their thought is that in the event of fire the individuals will see that the exits are chained shut and will automatically proceed to another exit just as they would change ticket window locations for quicker service. This thinking is fallacious because people are most likely to panic when they are forced into a locked exit, even though another exit is located nearby. This is especially true if real or supposed danger is immediately at hand.

Exits must be considered in the original construction of the building. Often, occupancy changes bring changes in exit requirements. Through periodic inspections, these changes can be noted by appropriate fire officials and proper remedial steps can be taken.

Proper maintenance of exits means that all exits are properly marked and usable at times when their use might be anticipated. The exit lights must be on, the exit doors readily operable, and all readily accessible without blockage of any kind. There is a tendency to place ticket booths, tables, soft-drink machines, and other objects in exit passageways without consideration of safety.

Public Address System. There should also be a means of obtaining immediate response from occupants to any emergency condition that might arise. Public address systems are extremely helpful. In some cases a public address system, operated by automatic control, has been used to replace the more normal fire bell or horn. Recordings over public address systems have been used to alert occupants to the existence of an emergency. It would seem that this concept will be used in future years. While a bell or horn is often used to signal an emergency, this sound does not always indicate an emergency condition to all occupants. Bells indicate telephone messages, changes of shifts, the need for a floor walker, as well as other nonemergency conditions. Public assembly occupancies may be equipped with both fire alarm and public address systems. This is a most desirable situation. Emergency power is essential to successful use of public address systems for fire evacuations.

Advance Notice to Patrons. In an effort to prepare people for the possibility of a fire or other emergency within the premises, some jurisdictions require advance notice to patrons to encourage them to look for exits within the structure. Jurisdictions may mandate the use of a printed program, printed sign, oral announcement, or showing on screen to the effect: "Notice—for your own safety look for the exits. In case of emergency, walk do not run to that exit."

Famous Panic Situations

Panic is not confined to fire situations. In May 1883, for example, the Brooklyn Bridge, spanning the East River between Manhattan and Brooklyn in New York City, opened for the first time. It was considered a great engineering feat of its day; however, many citizens were skeptical and did not feel that the bridge would stand for long. They were under the impression that this was just too great a feat for the human mind to accomplish. A week after it opened, many people were still apprehensive of the continued stability of the bridge. At that time a woman, in attempting to negotiate the steps of the Manhattan approach, fell down and screamed. Her screams caused many other people to think that the bridge was collapsing. In the panic that ensued 12 people were killed, and scores were injured. Some of those killed jumped off the bridge into the East River thinking that the bridge was collapsing. Actually there was no damage to the bridge whatsoever.[3]

In a fire-related panic situation, 70 boys and girls were killed in the Glen Cinema in Paisley, Scotland. The local History Library in Paisley recorded the incident as follows:

> In the disaster of 31st December 1929, 70 children aged between 18 months and 13 years died. In a room near the Gilmour Street entrance of the cinema (formerly The Good Templar Hall), a film had caught fire, sending up a cloud of thick black smoke which carried along a corridor into the main hall. On seeing this and thinking the building was on fire, the children rushed in a panic towards the exit at the opposite end of the building alongside the River Cart, only to find the exit doors were locked to prevent entry without payment. On arriving the fire brigade soon dealt with the small blaze and discovered the horrific scene; in their wild rush towards the exit some of the children had been trampled on, leaving piles of dead and dying children.[4]

More recently, panic occurred on January 29, 1956, when fire struck the Arundel Park Hall in Brooklyn Park, Maryland. At the time of this fire, approximately 1,100 people were in the hall. The fire broke out at approximately 5 P.M., and many of the patrons attending this church oyster roast believed that there was little danger; in fact the fire was taken as a joke by many occupants.

Dr. John L. Bryan, then head of the Department of Fire Protection Engineering, University of Maryland, conducted an extensive investigation of panic and other factors related to the incident. Twelve patrons were killed in the fire. In addition, 250 people were injured.[5] A quote from Dr. Bryan's report discusses panic conditions:

> As the heat, flames and choking smoke began enveloping the hall the patrons grew more panicky and began to push toward several exits. Suddenly the lights were extinguished plunging the hall into darkness leaving only the red glow of flames overhead with the heavy smoke over all. This is when panic prevailed and people fought for exits and many were knocked down or lost their footing and were trampled. Many of the persons who had fallen to the floor were picked up and carried from the building. Fallen chairs and overturned tables from which all glassware had been knocked added to the confusion and the injuries.[6]

Dr. Bryan's investigation revealed that trained individuals who happened to be in the hall, including police officers and fire service personnel, reacted very much the same as untrained individuals. He further found that many people did not leave by the same door that they had entered, although this is commonly thought to be normal procedure. At least 12 people were found to have safely exited from the building only to reenter.

Guard or Watch Duty Responsibilities _____

Within a given institution, mercantile establishment, or industry, the personnel on guard or watch duty may play a major part in the fire procedure program. The security officer is the only person on the premises during a majority of the hours in many occupancies, including mercantile establishments and office buildings. The security officer may therefore have the responsibility of discovering the fire, reporting it, and carrying out fire control duties. This individual therefore may constitute the weakest link in the entire operation of the facility. Unfortunately, this job is not always held by people who are capable of reacting adequately to emergency conditions.

Knowledge of Facility

The security officer should be made fully familiar with the entire facility. The purpose of all valves and drains, the piping system in the structure, and the location and operation of fire alarm boxes or other devices necessary in reporting a fire should be explained. It may be desirable to have the guard become familiar with the operation of the stationary fire pump in the plant.

Reporting a Fire

It is important to explain what to do in case of fire, where fire equipment is located, and how to operate it. The security officer should be taught how to summon the fire department and instructed in conditions under which to call for assistance. Unfortunately, some companies instruct security officers to refrain from calling the fire department unless a major fire is under way. Some of the most serious losses in North America have resulted from such a policy.

The security officer should be instructed in how to use first-aid fire extinguishers and understand that a fire extinguisher should be used strictly as a first-aid fire fighting device.

The importance of calling for assistance cannot be overemphasized. The security officer must be advised to call without delay and to take no chances of being overcome by smoke while trying to handle a fire. There may be a tendency on the part of some security officers to play the hero. This desire may even lead them to start fires for publicity purposes. Individuals employed as security officers should be screened thoroughly and a police record check made in every case.

General Duties

Guard duty responsibilities include discovering fire hazards, such as flammable-liquid spills, and reporting impaired fire protection equipment. The security officer might observe a shut sprinkler valve or a deactivated fixed fire extinguishing system or may discover the flow of water from a broken sprinkler line or an open sprinkler head and take appropriate action to remedy the condition.

On occasion, employees leave the premises without checking to make sure that all machinery has been shut down. The security officer can be of great help in detecting machinery left operating and can be responsible for either shutting it down or obtaining help in doing so. The officer may also detect breaks in steam lines.

Making the Rounds

Rounds should be make on a scheduled basis. This may include a round each hour with the trip recorded by a time clock with punch stations. The punch stations should be located in appropriate areas of the facility to ensure complete coverage of the premises.

Some facilities have an electrical system that transmits a signal to a central station if the security officer fails to insert a key in a given signal station according to the time scheduled for the appointed rounds. A runner then responds to see what has gone wrong.

Private Fire Brigades _____

Another means of ensuring proper fire reaction is through the organization of private fire brigades. Fire brigades are especially useful in industry and in institutions.

A fire brigade is organized to take immediate action in the event of fire. This action is of a first-aid nature and normally would only supplement, rather than take the place of, action by the public fire department. The public fire department may at times be delayed because of traffic conditions, slow transmission of alarms, inclement weather, bad roads, or simultaneous alarms. There are also occasions when the local public fire department is not capable of handling major fires in large industrial plants.

In a few areas, the private fire brigade is essential because there is no public fire department. This is true in some unincorporated areas that have resisted annexation by the nearby city or town.

The private fire brigade has an advantage over the public fire service in handling fires in industrial plants. This advantage results from the brigade's superior knowledge of the plant.

There should be brigade members in various sections of the plant or institution. Some enterprises use only maintenance personnel for fire brigades. As can be seen, this system has a disadvantage.

Fire brigade members have a primary responsibility to turn in an alarm on discovery of a fire. Except in a very large complex, this alarm will go to the public fire department as well as to other areas of the plant. In larger plants the alarm would go to the fire station located within the plant.

In addition to turning in alarms, the plant fire brigade has a responsibility for operating first-aid fire extinguishers and hose lines. Brigade members must be fully trained in the use of all extinguishers that they may be expected to operate. They must also understand the operation of automatic sprinkler systems.

Fire brigade members also serve as fire inspectors in their work areas. Trained brigade members realize the need for all sprinkler valves being open. They also recognize problems inherent with excessive accumulations of trash and blocked fire equipment.

Fire brigade members recognize that fire doors must be maintained in such condition as to permit closing in the event of fire.

The fire brigade has a responsibility to transmit an alarm to the public fire department and to direct it to the fire location. Fire brigade members also have a responsibility to close fire doors. The fire brigade should learn which doors need to be closed in the event of fire in any given location. Fire brigade members should also check to ensure that all sprinkler valves are open at the time of a fire and may have the responsibility for seeing that fire pumps are started and are operating properly.

A good fire brigade also performs salvage duties. Loss in a given fire can be materially reduced if personnel are available for salvage operations.

As previously mentioned, some large plants and institutions maintain full-time fire departments. These departments function in the fire safety inspection field in a manner similar to that of the public fire department. However, plant fire department members generally spend a considerably greater percentage of their time on inspection and fire safety education duties than would be the case in a public fire department.

The size of a plant fire brigade will depend on the size and area of the plant as well as on the nature of processes carried out within it. As an example, a plant that handles flammable liquids in quantities will require more fire brigade members than would a plant handling sheet metal. Another factor to consider in determining the size of the fire brigade is the location of the plant. A rural plant will require more brigade members than would one in an urban area. This is especially true if the rural area does not have public fire protection facilities adequate for anticipated fires in local industry.

Another factor for consideration is coverage of all shifts. Many plants have good fire brigades during the daytime hours, but their brigades do not operate on later shifts. The provision of backup personnel for vacations and sick leave must also be considered. Because of limited training time available, people with previous training and experience in fire protection should be used when possible.

The individual selected as chief of the fire brigade should be from the supervisory ranks of industry. An awkward situation is created when a nonsupervisior is selected as fire brigade chief. In some plants, the fire brigade chief is selected by management, while in others the chief is elected by vote of the brigade members.

The fire brigade should include some members from the maintenance force of the plant. Pipe fitters, electricians, and other maintenance employees can be highly valuable to a plant fire brigade. In some industrial plants and institutions, the entire brigade is made up of such personnel. By virtue of their primary responsibilities, maintenance people must take part in any fire suppression activity in the plant. It is not desirable to depend entirely on maintenance personnel for fire brigade duties.

Long after the fire is out, fire brigade members may find themselves standing by to control a rekindle. Such standby duty may involve maintaining a charged hose line near the fire.

Fire Brigade Drills

Brigade drills should be conducted to ensure continued efficiency of operation. To be most effective, drills should be scheduled on a monthly basis. A great deal of valuable time and effort is lost by holding drills when no imagination is used in planning. Some drills are set up using repetitive procedures that have been practiced time and time again. It is most important to pose problems that require some thinking on the part of the participants. All too often the drill is merely held to meet a requirement imposed by an insurance carrier, and the main objective is to be able to honestly report that a drill has, in fact, been held.

Drills held by industrial fire brigades, as well as those in institutions and mercantile establishments, should include occasional sessions in which public fire departments cooperate in prefire planing. It is also helpful to have fire service members visit industrial plants to give them a complete picture of plant fire protection features.

A fire bill or operating procedure should be prominently posted to specify the duties and responsibilities of individuals in the plant or institution. The posting of such a bill is especially desirable in an institution where considerable reliance is placed on the activity of nurses, orderlies, and other personnel who are not specifically designated as fire brigade members. The fire brigade program should not merely function on paper. The fire bill should be viable and should represent practical measures that can be carried out without a great deal of difficulty.

One of the major reasons for organizing a fire brigade is to have fire suppression capability on the premises. Some industries have no fire brigades because the attitude prevails that the fire department is so superior that there is no need for a plant fire brigade. In some cities, in fact, public fire officials have not encouraged the development of fire brigades because they assume their efforts would be sufficient to handle any eventuality. A number of conditions can arise that limit the effectiveness of even the best fire department in bringing a plant fire under control. The caliber of the local fire department should be of some concern in developing a brigade; however, it cannot be the sole determining criterion.

As an added factor, there is a problem today with high stacking of combustible materials. There is also a trend toward windowless buildings, which makes the job of fire suppression more difficult. Another fire problem that did not exist until recent years is the wide use of air-conditioning systems. These systems can be responsible for the rapid spread of smoke and fire throughout the premises, often in a manner that is not readily detected by fire suppression personnel.

Regulations of the U.S. Department of Labor's Occupational Safety and Health Administration relating to private fire brigades have discouraged some industries from establishing such units. OSHA requires training and proper protective equipment for individuals assigned to a brigade as part of their employment. Some industrial managers find these requirements so arduous that they decide to depend on public fire services rather than go to the trouble and expense of organizing a fire brigade.

Fire Safety Manuals

In many occupancies a formal fire brigade is impractical. In those locations a fire safety manual may be used to promote proper fire reaction. These manuals form the basis of fire drill procedures and are tailored to the specific occupancy in which they will be used.

An example of the well-planned fire safety manual is the one in use in the residence halls at Oklahoma State University, Stillwater.[7] The manual contains information on fire exit drills, fire prevention and inspections, regulations for decorations, emergency lighting, trash chute fire problems, fire procedures, fire control, fire extinguishers, fire alarm systems, false alarms, fire bombings, fireworks, general safety, and first aid. A typical floor plan is included for use in developing fire evacuation plans.

Public Elevator Fire Safety

Many jurisdictions require the posting of signs near passenger elevators to warn the public against using elevators in the event of an emergency. An example of this fire reaction program is the requirement that Maryland established in 1973, which requires a notice in all elevators: "WARNING: Elevator shall not be used in the event of fire. Use marked exit stairways."

The requirement was imposed as a result of several incidents in which individuals were killed or injured while attempting to leave burning buildings on elevators. The use of heat-sensitive door-opening devices has made the problem more acute because of the possibility that doors will open at the fire floor. At times it may be necessary to use an elevator to evacuate invalids or handicapped people in the event of an emergency, and such a regulation does not preclude this possibility.

Fire Reaction Safeguards in Housing for the Elderly

An outstanding example of guidance for instilling proper reaction to fire is the publication *Life Safety from Fire—a Guide for Housing for the Elderly,* a report prepared for the Architectural Standards Division,

Federal Housing Administration of the U.S. Department of Housing and Urban Development.[8] The report is specifically designed to pinpoint problems in housing for the elderly (age 65 and over). It was based on a study of thousands of fires, which were reduced to a detailed study of 1,000 fires. It points out that many elderly people are housed in buildings that were not specifically designed for them. It mentions problems in attempting to classify occupancy on the basis of known fire causes and in attempting to develop new classifications. New occupancy classifications are generally developed through experience with disastrous fires. For example, 48 nursing home fires resulted in 428 fatalities between the years 1951 and 1966. During this period one state after another took steps to provide more adequate fire requirements for this type of occupancy. As previously noted, the federal government imposed fire safety provisions soon thereafter.

Few, if any, jurisdictions provide specific requirements for apartment houses occupied by the elderly. The report suggests that a series of fires may need to occur before such requirements are imposed. The report emphasizes that elderly people do not react in the same manner as younger people. A code requirement for 100-foot travel distances to exits is mentioned as being satisfactory for younger people who might be able to run that far under smoky conditions. However, this is an almost impossible feat for a person over 80. A 21-year-old subject could easily escape from a second-floor window, while this would be practically impossible for a 78-year-old person. An individual's reaction time is considered most critical. This is the time it takes the occupant to react to the fire and then to attempt to escape.

An interlocked chain of life safety operates in any occupied structure. Links of this chain are fire prevention, detection, alarm, escape or refuge, confinement, control, and extinguishment. All are necessary for complete protection. To ensure 100 percent life safety from fire is a theoretical ideal, and an effort must be made to make the links of the chain as strong as possible. Ignition temperatures and flame-spread characteristics of various materials are discussed in the HUD report. The flame spread of untreated red oak as the norm is rated at 100, while cotton fabric in folds has a flame spread rating of 1,600 to 2,500. Fuel contributed and smoke and gases developed are other matters of concern in the overall fire picture when human reaction is considered. The fire resistance of materials in which the hourly rating is a factor is also of concern. A point is made of the matter of fire spread, which may be based on fuel available for burning. The presence of flame and the effect of smoke and heat on the human body are factors in finding probable paths of escape. The reduced oxygen content as a result of burning also must be considered.

In connection with fire reaction, age, health status, and need for assistance in evacuation are important, as are factors such as mobility, hearing, vision, and sense of smell. Approximately 35 percent of the ambulant population over the age of 65 have trouble climbing stairs and performing other complex movements such as climbing through windows. A study by the U.S. Public Health Service showed that two-thirds of all people 80 and over have significant hearing problems. This, of course, would have an effect on hearing the fire alarm, the fire itself, or shouted directions. A great number of old people suffer from poor vision in dim light and a tendency to be blinded by fire and flame. They also have a poorer sense of smell, which may have a bearing on original detection of a fire.

Another factor that might affect reaction is disorientation. Older people sometimes tend to become confused and disoriented, which may cause them to be reluctant to use a given exit in the event of a fire.

Toleration of heat, smoke, and gases also decreases with advanced age. This means that there is less likelihood of success in escaping from a fire. Shock may also be a factor and may be closely related to heart failure, which is an ever-present hazard of advanced age. Psychological factors affect the overall fire situation. For example, old people typically collect objects and memorabilia in apartments and sometimes lose interest in keeping neat and orderly quarters. If a fire does break out, there is a possibility that the person detecting the fire will take no steps to sound an alarm or notify other individuals living in the structure. The person may go directly outside and away from the building without taking any steps to notify other people.

An elderly person may also be more inclined to attempt to put out the fire to maintain a sense of self-sufficiency. This attitude may be a matter of personal pride, and the person may feel that failure to control the fire may mean being forced to give up living independently and moving back with relatives or into an institution.

Many elderly people view the contents of their rooms or apartments as their only significant material possessions. As a result, the decision to leave, especially if the fire cannot be immediately seen, may be an extremely reluctant one. In a number of instances elderly people have refused to leave their rooms during fires and other emergencies and have had to be forcibly removed by fire fighters.

An elderly person may easily overlook such simple protective measures as using wet blankets and pretesting doors for heat. Closely related to this problem is the tendency to follow familiar routes in an emergency escape, which is especially characteristic of elderly people.

Reaction times for elderly people are usually almost twice as long as for young people. Specific designs of exit arrangements allow for decreased travel distance for elderly people. In considering places of refuge, there is the possibility that individuals—especially the elderly—should stay in their apartments, the safest place of refuge.

Compartmentation has been shown to be helpful in minimizing the danger of fire spread from one apartment to another.

For fire safety, it is important to limit combustibles in hazardous areas and to provide automatic sprinkler protection. Fire drills can often be effective, and they may even include response by fire department equipment on a nonemergency basis. The fire plan should give no consideration to use of fire extinguishing devices by occupants of the structure. Elderly occupants should not be assigned to fire control duties. Under no conditions, the report suggests, should the occupants be counted on to operate fire extinguishing devices but emphasis should be placed on these individuals sounding alarms and evacuating the building.

A detailed study such as this can be of great assistance in instilling proper public fire reaction. Many other occupancies have unique problems that should be addressed. This report may well serve as a guideline for future reports on occupancy classifications. It is interesting to note that the report considers psychological factors and that diverse reactions may be anticipated.

Fire Reaction for Special Groups _____

Major fires involving loss of life in facilities for mentally retarded and impaired people have focused interest on fire reaction programs for residents of such structures. The Massachusetts Firefighting Academy and the Department of Mental Health, with funding support from the Office of Special Education and Rehabilitation, U.S. Department of Education, prepared a trainer's guide entitled *A Fire Safety Program for Mentally Retarded Adults*. This manual is based on studies of fires in such facilities as well as research into the evacuation behavior of residents of typical facilities.[9]

This guide, which was distributed by the National Fire Protection Association, is designed for use by fire service personnel, facility operators, and state licensure personnel. It includes special concerns for individuals with hearing impairments and visual and physical handicaps.

Changing concepts of residential treatment create new challenges for those involved in developing proper fire reactions. As an example, several cities now have hotel-hospitals. This innovation enables people to move from a hospital, with its high daily room cost, to a adjacent hotel, which is operated by the hospital.

Patients can receive limited care in the hotel-hospital but can to be taken back into the hospital without delay if necessary. This is especially useful for individuals who live some distance from the hospital or who have no one at home to provide post-hospital care.

Reaction to fire may change materially in the hotel-hospital, because it usually has fewer staff than a hospital. This means that the patients must fend for themselves to a much greater extent. Code developers have difficulty in revising fire safety requirements rapidly enough to keep up with these use and occupancy changes.

Fire Reaction in High-Rise Structures _____

High-rise buildings present serious fire reaction problems. A number of factors may make it difficult for individuals to get out of these building in the event of fire.

Major problems develop because of the common use of the elevator for reaching all floors above ground floor. People occupying high-rise buildings seldom use stairways. This means that personnel employed in the building, as well as visitors, may not be familiar with the location of stairways. An example of this problem was encountered in a fire that occurred in a large southwestern city in a nine-story modern office building. This building was equipped with air-conditioning, and all windows were of the fixed-sash type so that the balance of the air-conditioning system would not be disturbed, a common procedure in structures of this type.

The fire occurred during working hours; it was of relatively minor consequence and was confined primarily to the basement of the structure. The problem was that smoke from the fire rapidly permeated the upper floors of the building through the air-conditioning system. The fire was brought under control, and fire fighting equipment was being prepared for return to service when a large executive chair came hurtling down from an upper floor. The chair hit the hose bed of one of the fire pumpers.

Personnel in the street looked up, only to see that occupants had gathered around a window that had been broken open by throwing the chair through it. The window opening was used to call for help. The aerial ladder, which was still on the scene, was raised to the floor level, and several people were rescued by ladder. Fire department personnel went up the stairways to the top floor and assisted other people to safety. The group of people assembled on the top floor were almost in a state of panic and had not given thought to using either of the stairways that were readily accessible nearby. For some reason, no one thought of trying any means of egress other than the elevator, which was customarily used.

The problem occurred primarily because the window could not be readily opened. Countless high-rise buildings throughout North America present similar evacuation problems. Normal fire reactions must often be modified in such structures. It is quite difficult to have occupants walk down the stairs for a number of stories to reach a place of safety.

Many studies have been carried out to develop improved fire safety procedures for high-rise structures. One of the most promising developments was the voice communication system for advising people about fire conditions. It has been thought that bell or horn alarms were adequate; however, in many conditions the voice alarm would be far more desirable, at least as a supplement to the fire alarm system.

Protection for high-rise structures should include automatic sprinkler protection. With this equipment normal fire reactions might be entirely suitable, because fires would be held to such minimal proportions as to enable egress by normal methods.

As with any other type of structure, the high-rise should have specific fire protection procedures and requirements. If elderly people are housed at an upper level, fire protection requirements will be somewhat different from those for normal office occupancy.

Lighting is another special factor for consideration in the high-rise occupancy. A power failure at a time of darkness can be extremely critical with or without a fire. The well-publicized power failure in the northeastern section of the country in 1969 brought special emphasis to this particular problem. The high-rise should be equipped with emergency lighting units or emergency generators to give occupants the opportunity to reach a place of safety without appreciable delay. They should not have to stumble down many levels in the dark to reach the outside of the building. This problem was re-emphasized in the World Trade Center bombing in New York in 1993.

A study by the Travelers Insurance Company indicated that full evacuation of certain buildings during fires is not only unwarranted but unwise. This study of incidents in hotels, apartments, and other medium and high-rise buildings showed that occupants are more likely to survive by staying in their rooms when the building is of fire-resistive construction and is equipped with fire safety devices.

As an example of the application of this principle, building operators in Gainesville, Florida, have been given the option of permitting occupants to stay in place (except when their rooms are on fire) if the following conditions prevail: the building is of fire-resistive construction; fire procedure instruction cards are placed in each room; a public address system with emergency power is available to alert all occupants; each living unit is equipped with electrically connected smoke detectors; the building has full automatic sprinkler protection; at least one window is openable or breakable in each living unit; individual heating, ventilation, and air-conditioning units are provided, thereby limiting vertical and horizontal smoke spread; a complete fire alarm system is provided; one-hour fire separation is provided between living units and corridors; quarterly in-house fire inspections are conducted;

at least two building employees are on duty at all times; and the fire alarm system is connected to the fire department

An evacuation drill in a high-rise apartment building for the elderly in a major city produced two cardiac arrests plus other disabling injuries. The city in which this occurred developed a modified or limited evacuation procedure in view of this experience. As a greater percentage of high-rise buildings are equipped with superior fire protection, the nonevacuation procedure will likely gain acceptance.

Summary

An integral part of fire prevention planning and education is instilling an awareness on the part of the public on how to act and what to do in the event of a fire, at home or in public places. Any assessment of human reactions that might be anticipated in a fire emergency must take into account individuals of all ages, health classifications, physiques, occupations, and temperaments. Fire severity, location, the manner in which it started, and the presence of the individuals in the fire area must also be considered.

One of the first things to stress is the importance of summoning the fire department promptly by going to the nearest telephone or fire alarm box and giving the fire department dispatcher accurate information about the location of the fire. Stickers on telephones, signs on fire alarm boxes and on the sides of fire department vehicles—even billboards—can be used to keep the fire department telephone number constantly before the public.

Fire drill practices in homes, schools, institutions, and industry are an effective means of teaching people how to respond in a fire emergency in a planned, orderly way. Fire drill programs in schools throughout the United States and Canada have probably had a more lasting effect in conditioning the public to proper fire reaction than any other single effort.

There is a need for more widespread public education in the proper use of portable fire extinguishers and particularly in recognizing their limitations, both in the home and in public buildings. Although fire extinguishers have been used successfully on thousands of fires, many fires involving large losses have ensued because of failure to use fire extinguishers properly or delays in reporting a fire while attempting to use an extinguisher.

In planning measures for improved public fire reaction, the planner must consider conditions that cause panic and steps to take to counteract panic, especially in public places. Panic is easily triggered in a crowd, but certain physical measures can help to prevent it: enforcement of maximum occupancy limits; emergency lighting; clear designation and maintenance of exit facilities; use of public address systems for emergency instructions; and advance notice to patrons to look for exits nearest them to be prepared in case of fire. Closely allied to those measures is posting of signs near public elevators warning passengers not to use the elevators in the event of a fire and to use marked exit stairways instead.

Fire reaction safeguards in housing for the elderly and in high-rise structures have received special emphasis as a result of several serious fires. Studies are being made of the special problems in both types of occupancies. As one report puts it, the essential links in the interlocked chain of life safety are measures for fire prevention, detection, alarm, confinement, control and extinguishment of the fire, and escape or refuge from the fire.

Not to be overlooked is the role of the guard or watch personnel in fire protection of the premises in institutions, mercantile establishments, and industrial plants. Care should be taken to hire responsible people for this job and to give them adequate training to ensure proper fire reaction.

Private fire brigades, maintained by some institutions and industrial plants and trained in fire reaction procedures, have contributed immeasurably to public fire safety. Though intended primarily to supplement action by the public fire department, fire brigades can be critically important when the public fire department is delayed in answering the alarm or when it is overburdened during periods of civil disorder or natural disasters, or in areas where no public fire service is available.

Notes _____

1. Bernard M. Levin, unpublished paper presented at National Fire Protection Association Fall Meeting, Nashville, Tenn., 1988.
2. *Life Safety Code,* NFPA Standard 101 (Quincy, Mass.: National Fire Protection Association, 1994).
3. John L. Bryan, "Psychology of Panic," paper presented at Fire Department Instructors' Conference, Memphis, Tenn., February 18–19, 1958.
4. Personal communication from Ann E. Martindale, Local History Library, Central Library, Paisley, Scotland.
5. John L. Bryan, *A Study of the Survivors' Reports on the Panic in the Fire at the Arundel Park Hall in Brooklyn, Maryland, on January 29, 1956* (College Park: University of Maryland, 1956).
6. Ibid., p. 6.
7. *Fire Safety Manual,* Oklahoma State University Residence Halls (Stillwater: Department of Fire Safety, Oklahoma State University).

8. U.S. Department of Housing and Urban Development, *Life Safety from Fire—a Guide for Housing the Elderly,* report prepared for the Architectural Standards Division, Federal Housing Administration, by Raymond D. Caravaty and David S. Haviland of the Center for Architectural Research, Rensselaer Polytechnic Institute, Troy, N.Y. (Washington, D.C.: U.S. Government Printing Office).

9. Massachusetts Firefighting Academy and Department of Education, *Trainers Guide: A Fire Safety Program for Mentally Retarded Adults* (Boston).

CHAPTER 9

Fire Prevention Efforts of the Private Sector

Visitors from foreign countries who are in the United States and Canada studying fire protection activities are often amazed at the important role played by nonpublic fire safety agencies. In many countries, the private sector is of little significance in the development and implementation of fire protection procedures. By tradition, fire prevention activities in the United States and Canada have been motivated to a great extent by private organizations.

Fire safety activities of the private sector may be divided into three categories: those in which fire prevention is the primary function of the nonpublic organization; those in which fire prevention is a secondary effect of activities in the promotion of trade; and those in which fire prevention activities are carried out for the internal protection and well-being of the organization.

Organizations with Primary Fire Prevention Functions

National Fire Protection Association

In the category of organizations having the prevention of fire as a primary responsibility, there would be little doubt that the National Fire Protection Association (NFPA), headquartered in Quincy, Massachusetts, would hold the top position. The objectives of the NFPA, as stated in its articles of association, are "to promote the science and improve the methods of fire protection and prevention; to obtain and circulate information on these subjects; and to secure the cooperation of its members and the public in establishing proper safeguards against loss of life and property by fire." The Association was founded in 1896.

The organization is a nonprofit educational and technical association devoted entirely to preventing loss of life and property by fire. Membership is open to any individual, company, or organization having an interest in these objectives. Although most members are from the United States and Canada, the NFPA has worldwide impact in the fire prevention field, as well as members from more than 70 countries.

Activities of the NFPA include the developing and publishing of technical standards, codes, recommended practices, and model ordinances. More than 285 such publications have been developed, prepared, and updated under the guidance of 210 NFPA technical committees. More than 5,200 individuals serve on these committees, all on a volunteer basis. These technical committees usually include representatives from industry, public fire services, manufacturers of equipment, the insurance field, governmental agencies at all levels, and other qualified individuals. Committees are arranged to preclude their domination by any represented groups that may have special interests in the subject at hand.

The standards developed are prepared only after detailed study and vote by the committee. Built into the process of developing and adopting standards is the publication of calls for proposals to amend existing documents or to shape the content of new documents. These public proposals, together with the committee action on each proposal, as well as committee-generated proposals, are published in the Report on Proposals for public review and comment. Public comments, along with the committee action on each comment, are then published in the Report on Comments. Only after this public review and comment cycle has been completed is the final committee report brought before the membership for action. Once adopted by the NFPA membership at either an annual or a fall meeting and issued by the Standards Council, standards are published and made available for voluntary adoption. Many of these codes and standards have great influence and are used as the basis for legislation and regulation at all levels of government.

The NFPA has for many years published educational materials in the fire prevention field. These include standards, pamphlets, posters, occupancy studies, fire records, films, videos, and a wide variety of materials suitable for use in local fire prevention promotional efforts. More than 800 publications are available from the Association. The NFPA also maintains the world's most comprehensive fire protection library at its headquarters.

The NFPA provides technical and other topical fire safety information in its membership magazine entitled *NFPA Journal,* which is published six times a year. Other periodical publications for members include *Fire News,* a bimonthly newsletter containing current fire protection information and standards-related articles.

Among the many books published by the NFPA, the one best known is undoubtedly the *Fire Protection Handbook.* The first edition appeared in 1896, and through the years it has become the bible of the fire protection field. Other handbooks include *Industrial Fire Hazards, Automatic Sprinkler Systems, National Electrical Code, Life Safety Code,* and *Flammable and Combustible Liquids Code.*

The NFPA is the leader in public education programs concerning fire protection and prevention. "Learn Not to Burn" is the theme and focus of the NFPA's comprehensive public fire safety education program, which includes a national media campaign, a fire safety curriculum for school children, and an extensive outreach program supported by a regional field network of NFPA representatives. Fire Prevention Week is sponsored annually by the NFPA as a major national event designed to promote fire safety awareness. Sparky ©, the Fire Dog, is the creation of the NFPA.

The NFPA's Engineering Services Division has several field service projects working in specific phases of fire protection, including electrical, life safety, flammable liquids, gases, and marine fire protection. These special field service groups provide guidance in the application of pertinent standards in specialized areas.

The Association annually publishes statistics on the fire problem in the United States and maintains one of the world's most extensive fire experience data bases. Combining its annual survey of fire departments and Fire Incidence Data Organization (FIDO) file on large and other fires of major technical interest, the NFPA publishes two dozen major annual overview reports on the nation's fire problem. It also provides numerous special topical reports and responds to thousands of individual and organization requests for data.

The Fire Investigations Department conducts investigations of major fires of technical or educational interest. Important lessons learned from the fires investigated provide input to NFPA technical committees and are vital in understanding how fires originate and how similar fires can be avoided in the future.

In 1982, the NFPA formed the National Fire Protection Research Foundation with the specific goal of identifying fire problem areas and developing solutions to those problems. Foundation programs help meet the urgent need to begin or continue research into hazardous materials, improved safety equipment for fire fighters, better and more realistic code enforcement, improved public education, and a safer environment.

The NFPA has more than 63,000 members. Many of them are affiliated with one of the 14 membership sections. These include the Fire Marshals Association of North America, the Industrial Fire Protection Section, the Fire Service Section, the Electrical Section, and the Railroad Section. Each section has elected officers and carries out functions relating to fire prevention within its sphere of interest. Business meetings of these groups are normally held during the NFPA annual and fall meetings.

In an effort to provide better service to members, the NFPA has branch offices in the Washington, D.C., area, and Ontario, California. The Washington office serves as a liaison with all federal agencies on many matters in the fire prevention field. With the ever-increasing interest of the federal government in fire protection, coordination between federal and private agencies has become imperative. Both NFPA offices have been valuable in coordinating efforts and gaining recognition of the part played by nongovernmental interests in fire protection.

Other Fire Prevention–Oriented Organizations

The International Association of Fire Chiefs has as a primary objective the encouragement of fire prevention activities by local fire departments. The organization has for many years publicized and promoted a home inspection program and other fire safety programs in the interest of reducing the number of fires in dwellings. The association is composed primarily of fire chiefs and their deputies in the United States and Canada.

The International Association of Arson Investigators is an organization of professional fire investigators. Its primary objective is the prevention of fire through the suppression of arson. The association encourages adequate training to enable its membership to fulfill responsibilities in this important field.

The International Association of Fire Fighters, an affiliate of the AFL-CIO, represents a large percentage of career fire department personnel in the United States and Canada. Local members have been active in fire prevention and have on occasion directly sponsored fire prevention displays at public gatherings.

The Canadian Association of Fire Chiefs is also quite active in fire prevention efforts. It is not affiliated with the International Association of Fire Chiefs in the United States. In addition to direct sponsorship of many fire prevention programs, the Association is a major sponsor of Fire Prevention Canada.

Fire Prevention Canada has been successful in introducing a number of national fire prevention programs in Canada. Through the support of the Canadian Child Protection Foundation, "Learn Not to Burn" is being implemented throughout Canada. Both French and English language versions are used.

The organization is also supported by the Insurance Bureau of Canada. Another major sponsor is the Association of Canadian Fire Marshals and Fire Commissioners. This association comprises the provincial fire marshal or fire commissioner from each of the Canadian provinces and territories as well as the military and government fire marshals. This is a most prestigious group and has major impact on Canadian fire safety practices.

The National Safety Council, an organization having major impact in the field of safety, includes fire prevention as an overall safety objective. The annual National Safety Congress is probably the largest gathering in the world of people with an interest in safety. Many of those active in the Council have fire prevention responsibilities as part of their overall safety duties. Canada Safety Council has a similar role north of the border.

The Society of Fire Protection Engineers, which started as an arm of the NFPA and is now a separate organization, serves the professional field of fire protection engineering. The development and encouragement of fire prevention are major activities of the society.

The International Association of Black Professional Fire Fighters was formed to foster liaison among black fire fighters. The organization has an interest in the promotion of fire prevention in high fire risk areas.

The International Fire Service Training Association, formed in 1943, develops and publishes training material for fire service personnel. Manuals on fire prevention subjects are included. Committee members from throughout the United States and Canada gather annually at Oklahoma State University to update the manuals and develop new material. A majority of states and provinces use these manuals for their fire service training programs.

The International Society of Fire Service Instructors was organized in 1960 by a group of state and provincial fire service instructors. The society has since expanded to include local training officers as well as other fire service personnel. The society has strived for improved instructor capabilities and has generally represented fire service training interests at the national level. Improved training in fire prevention is a goal of the organization, which has formed the Alliance for Fire and Emergency Management.

The National Volunteer Fire Council represents the interests of the nation's volunteer fire service. Many volunteer fire departments have aggressive fire prevention programs.

The National Association of State Fire Marshals represents state fire marshals at the national level. This organization has actively supported fire safety education, smoke detector installation, and other abatement programs.

Organizations with Allied Interests in Fire Prevention

Private sector organizations having fire prevention as an adjunct of their operations may be divided into several categories. These include professional associations that issue standards having some bearing on fire protection; trade associations that deal in areas having some relationship to fire protection; organizations that develop model building codes; and the insurance industry.

Professional Associations

Examples of the associations that issue standards having a bearing on fire protection include the Air Conditioning and Refrigeration Institute, the American Industrial Hygiene Association, the American Institute of Architects, the American Society for Testing and Materials, the American Society of Agricultural Engineers, the American Society of Heating, Refrigerating, and Air Conditioning Engineers, Inc., the American Society of Mechanical Engineers, the American Society of Safety Engineers, the American Water Works Association, the American Welding Society, the Institute of Electrical and Electronics Engineers, and the NFPA. All of these groups prepare standards and procedures that have some bearing on fire protection.

The American National Standards Institute does not prepare or write standards; rather, standards are prepared by some 250 separate standards developers, including those listed above. Many of these standards are then submitted for approval as American National Standards (ANS). Therefore, the American National Standards Institute is the coordinator and issuer of ANS designations.

Trade Associations

Trade associations that have some interests bearing on fire prevention include the American Fire Sprinkler Association, the American Forest and Paper Association, the American Hospital Association, the American Health Care Association, the American Gas Association, the American Iron and Steel Institute, the American Petroleum Institute, the Automatic Fire Alarm Association, the Compressed Gas Association, the Fire Equipment Manufacturers Association, the Gypsum Association, the Institute of Makers of Explosives, the Lightning Protection Institute, the Manufacturing Chemists Association, the Manufactured Housing Institute, the National Association of Fire Equipment Distributors, the National Association of Home Builders, the National Electrical Manufacturers Association, the National Fire Sprinkler Association, the National Propane Gas Association, and the Society of the Plastics Industry. These trade associations, as well as a number of others, have an interest in products and services that relate in some manner to fire safety. They often include fire prevention subjects in their publications and in programs presented for their membership.

To illustrate how a trade association may be concerned with fire prevention, the National Propane Gas Association will be reviewed in more detail. The industry represented by the association markets a product that is inherently hazardous. To be marketable, the material must have the ability to burn—a characteristic that creates a problem in handling the material. To minimize dangers, the association has taken steps to be sure that all individuals who handle the material are properly indoctrinated in safety measures.

The National Propane Gas Association, with headquarters in Chicago, has maintained active representation on the Liquefied Petroleum Gas Committee of the NFPA. They have participated in the development of standards for the safe use and handling of gas and have actively promoted employee training for member organizations. Liquefied petroleum gas associations at the state level are primarily concerned, as is the national organization, with the development of industrial and public uses of liquefied petroleum gas, but they also place considerable emphasis on safety programs of member companies. Contrary to popular opinions, all these associations have actively worked to promulgate and enforce liquefied petroleum gas safety codes in many jurisdictions.

The American Gas Association, in addition to performing trade association services, operates testing laboratories in Cleveland and in Los Angeles for the purpose of testing all types of gas appliances. Safety testing is a major part of this function.

Trade associations operating entirely in Canada that have similar interests include the Canadian Automatic Sprinkler Association and the Canadian Gas Association. There are also worldwide organizations welding together several of the national trade groups mentioned.

Building Code Groups

Building code development groups have a major impact on fire prevention. The building codes in effect in various jurisdictions throughout the United States and Canada play a major part in establishing fire protection requirements from a structural standpoint.

Building code groups operating at the national level in the United States include the Building Officials and Code Administrators, Inc., The International Conference of Building Officials, and the Southern Building Code Congress International. In Canada, building codes are developed by committees of the National Research Council of Canada, a government agency.

In an effort to bring about greater uniformity among the three model building codes, the Board for Coordination of Model Codes meets to work toward reducing conflicts among codes. This group comprises representatives of the model code organizations and the NFPA.

The Council of American Building Officials represents the model code groups. This council seeks to eliminate conflicts among codes. It sponsors a one- and two-family dwelling code.

In addition, the National Conference of States on Building Codes and Standards works toward uniformity of building regulations. The group comprises state government representatives as well as other classes of members.

Insurance Industry

As a result of the major fires that ravaged many American cities in the late 1800s and early 1900s, the property insurance industry played a major leadership role in the development an encouragement of fire prevention and fire suppression measures in the United States. Many advances came about as a result of the direct relationship between the fire insurance rate structure and the consequent implementation of local fire protection efforts.

In the 1980s the insurance industry's emphasis on fire protection began to shift. The progression of insurance policies from covering fire and lightning to package policies covering several property perils as well as liability lessened the insurance companies' centralized emphasis on the risk from fire alone.

In addition, strong competition developed among the insurance companies, which resulted in independent postures, lessening their joint efforts. The "mix" of business written by insurers has changed dramatically. Companies known as direct writers have concentrated their sales efforts on personal lines, whereas the large agency companies have emphasized commercial risks. Rather than supporting broad-based industry efforts, more and more companies maintain loss control departments specializing in the accounts each individual company wishes to insure.

There also have been major expense pressures on the insurance companies. As a result, many companies have reduced their contributions to trade organizations. That in turn has shifted those organizations' emphasis from public service activities, such as providing loss control efforts, to primarily dealing with federal and state government relations. With the reduction in fire protection activities by the insurance industry, more and more of the functions they formerly performed are now conducted by governmental, educational, and private nonprofit organizations.

Those insurance organizations currently performing major activities in the fire protection field follow:

Factory Mutual System. This organization represents three mutual fire insurance companies who specialize in coverage for highly protected industrial risks. The Factory Mutual System maintains a laboratory for the testing of fire protection equipment. The equipment meeting the system's standards is eligible to bear the FM seal of approval, which is nationally recognized. Corporations having Factory Mutual insurance coverage are surveyed on a periodic basis by system engineers for fire prevention purposes, as well as testing to ensure that all fire protection equipment is in proper operating condition.

Industrial Risk Insurers. The purpose of this organization of more than 40 property insurance companies is to insure well-protected industrial plants and institutions. The organization provides combined engineering and loss control surveys services to insured factories, hospitals, and other large occupancies. The services of its nationally recognized fire safety laboratory are available for training of personnel in the maintenance and use of various types of fire protection equipment normally found in a major industrial plant.

Insurance Services Office, Inc., and Its Wholly Owned Subsidiary, Insurance Services Office Commercial Risk Services. These two organizations perform the functions that before 1971 were performed by regional and individual state rating bureaus. These were functions such as insurance form and rate promulgations, in assigning a fire insurance grade to municipalities and fire districts, as well as rating individual commercial risks.

The function of Insurance Services Office, Inc., is to develop loss cost information, which is used by insurance companies to develop rates. It also promulgates standard policy forms and endorsements. Insurance Services Office, Inc., additionally serves as a statistical agent for subscribers desiring the obtaining of combined industry data.

Insurance Services Office Commercial Risk Services is responsible for surveying those risks that are specifically rated. It also is responsible for application of the Fire Suppression Grading Schedule. It also performs flood insurance surveys for the Federal Emergency Management Agency.

Insurance Services Office Commercial Risk Services conducts periodic surveys of communities in the United States to evaluate their fire protection facilities for the purpose of establishing a base grade. It is from this grade that property insurance rates are established. Field rating representatives assigned to this function use the Fire Suppression Rating Schedule to determine a community's classification. The schedule evaluates a community's water supply and fire suppression abilities. The largest jurisdictions in the United States are no longer graded using the Fire Suppression Rating Schedule, but rather, their own loss experience is used in developing their specific loss cost figures.

Results of municipal surveys, which are presented in the grading report, and the subsequent application of an insurance rate structure, had a profound effect on the establishment of fire prevention bureaus in many municipalities. Former leaders of many fire departments have indicated that the fire prevention bureaus were initially established only in an effort to satisfy the requirements of the insurance industry's grading schedule.

Rates for buildings and contents are generally established by the Insurance Services Office Commercial Risk Services staff operating within each state. A handful of states have their own individual rating bureaus, which continue to perform those functions. Rates for an individual building are based on the type of building, including occupancy; the construction of the building; the existence of an automatic fire sprinkler protection system within the structure, as well as the water supply available for such protection; the base rate, which is in turn determined by the overall loss experience within each state; and the fire defenses of the community as determined by application of the Fire Suppression Rating Schedule.

In all states, insurance rates as established by Insurance Services Office, individual state rating bureaus, or directly by insurance companies are subject to oversight by the state insurance department. The insurance code in each state provides a basis for the insurance commissioner to challenge rate revisions.

Twenty-nine states, the District of Columbia, and Puerto Rico have FAIR plans for writing either basic property insurance or package policies. The acronym FAIR stands for Fair Access to Insurance Requirements, and the plans permit insurance coverage for properties that would otherwise find coverage difficult to obtain.

FAIR programs were developed primarily as a result of the urban riots in the late 1960s. In those states that have FAIR plans, all insurance companies licensed to do business in the state are required to participate. Each company is assessed to cover the plan's losses based on the percentage of business the company transacts in the state as compared with the total premiums written. FAIR's inspections are conducted on individual risks, and coverage may be denied if the risk of loss is considered too severe.

Underwriters Laboratories Inc.

Underwriters Laboratories Inc. (UL) is an independent not-for-profit organization having a comprehensive program of testing, listing, and reexamining equipment, devices, materials, and assemblies as they relate to fire protection and safety.

The laboratories, which were established in 1894, test building materials of all types, fire protection equipment, gas and oil equipment, accident and burglary prevention equipment, as well as electrical appliances and equipment. The electrical testing program of UL, as an example, has had a major impact on fire protection, especially through the reduction of fires caused by the faulty design and manufacturing of electrical devices and appurtenances. The UL mark is internationally recognized as indicating that minimum standards for safety have been met by the product. Comprehensive tests have been devised to reduce the possibility of fire, shock, or injury of any type through the use of a product listed by UL.

An important feature of UL is its follow-up service program. UL's field representatives visit the factories where products are produced under UL service at least four times annually. The public is able to discern which items have been tested because of the use of a printed designation on the product. Almost seven billion labels are applied annually.

Underwriters Laboratories writes most of the standards used to investigate products. Those standards are regularly reviewed and revised on the basis of such considerations as field experience that indicates that a product is unsatisfactory from a fire protection or safety standpoint. Product directories issued semiannually show names of manufacturers of products that are listed by UL.

The organization has established representatives in many countries outside the United States in addition to its laboratories at Northbrook, Illinois; Santa Clara, California; Melville, New York; and Research Triangle Park, North Carolina. Underwiters Laboratories engineering subsidiaries operate in Hong Kong, Taiwan, Japan, and elsewhere. Inspection centers are maintained worldwide in more than 70 countries. Over 900 engineers are included in a total of over 4,000 employees.

The Canadian Standards Association and Underwriters' Laboratories, Inc. of Canada provide somewhat similar services in Canada. Both organizations maintain testing laboratories.

As an example of public utilization of a privately sponsored and developed activity, several jurisdictions require that all electrical equipment sold within the state or municipality be listed by the UL or another recognized testing laboratory. Maine has had such a requirement for a number of years. This came about because this state, in the far northeastern corner of the United States, was becoming a dumping ground for inferior or faulty electrical equipment from all parts of the country. North Carolina enacted a similar requirement in 1934, while Maryland followed in 1973.

Private Organization Efforts in Fire Prevention _____

Fire prevention activities carried out by private organizations for the improvement and protection of their own properties are many and varied. In the majority of business establishments, industries, and privately operated institutions and schools, the primary concern with fire prevention is that brought about by the local fire department, fire marshal's office, and insurance carriers. Some of the more progressive businesses maintain extensive internal fire prevention programs.

One hotel and motel chain has full-time employees vested with fire safety responsibilities. These employees travel to all of the restaurants and hotels in the chain to ensure that personnel are proficient in fire prevention activities and that hazards are kept to a minimum. Proper operation of fixed fire protection equipment is also checked. This corporation, The Marriott Corporation, has made giant strides in installing sprinklers in their properties.

Industrial and institutional fire prevention may require more complex safeguards, such as the formation of fire brigades having fire prevention as one of their responsibilities. A well-trained fire brigade can be of considerable value in abating fire hazards.

Some industries handle fire prevention at a local level with a person or persons being assigned in each plant for such duties. In other cases a floating fire prevention person may service a number of plants out of a central location.

Self-Insurance Programs

An example of a program in which there is a direct relationship between fire prevention efforts of private enterprise and expenditures is the self-insurance program. Under a self-insurance program, a business organization or government body accepts the risk connected with fire, either on a partial or complete basis. The industry or government body elects to handle fire losses with its own resources, rather than depending on commercially available fire insurance and/or comprehensive liability insurance.

The amount of self-insurance may vary widely. One company might have self-insurance to cover all losses—even under $1,000—while another may make the break at $10,000, $100,000, or possibly up to $5 million. Another company might decide to obtain no commercial fire insurance whatsoever and thus self-insure its entire operation. As can be appreciated, insurance rates decrease with higher deductible limits of liability on the part of the insurance carrier. Some government entities, including states, counties, and municipalities, have elected to follow a self-insurance program.

The federal government of the United States might be considered a leading self-insurer. The federal government does not carry insurance on properties or their contents. It depends on replacement by direct appropriation for any losses incurred, rather than on maintenance of a reserve. Individual property owners associated with governmental operations do carry commercial insurance, however. An example would be an officer's club located on a military base. The club management might elect to carry fire insurance coverage on equipment owned and used therein.

A self-insurance program places a direct responsibility on management, whether the program is operated by a governmental agency or private organization, to be sure that all proper precautions are taken to avoid fire. In that there is a chance that losses might not be fully replaced or that the reserve fund might be depleted, an obligation exists to ensure minimal losses.

The most effective self-insurance plans are those in agencies or businesses that have a spread-risk arrangement so that the possibility of a total loss at one location is avoided. A state government, for example, has hundreds of fixed installations; the likelihood of a number of these installations having a major fire during a given year is highly remote. As long as the reserve to handle the anticipated losses can be maintained, the plan should function satisfactorily.

Reserve Fund Maintenance. One unfortunate feature of the self-insurance plan is the possibility of use of the reserve fund for other purposes. If, for example, a company with a self-insurance program was to decide to open a new operation requiring funding, there would be a real temptation to tap the fire-loss reserve for this purpose. A fund of several million dollars lying in readiness for use in the event of a fire loss might be highly tempting to hard-pressed officers in either private or public agencies for diversion to meet more meaningful immediate obligations. A low limit of depletion for the reserve fund must be set to ensure a successful program.

Many self-insurance programs exist at the state level. At least nine states have a funded self-insurance program in which a definite reserve is held aside for use in the event of fire loss. In some cases procedures exist for the carrying of commercial fire insurance for major losses. In 16 states a procedure is followed for direct payment of losses by the state government without the existence of a specific self-insurance program. Fire losses are merely paid from general funds.

Fire Safety Requirements. A self-insurance program must consider provision for enforcement of requirements necessary to maintain low fire losses. The usual procedure is to give someone or some group the responsibility for determining what measures must be taken to make the place reasonably safe from the standpoint of loss. In industry a consultant firm is often used to provide outside inspections for the company.

Property protection requirements under a self-insurance program are sometimes not the same as code requirements enforced by public fire officials. Authority for inspections might be set up in the administrative procedure for the self-insurance program. The industrial plant manager must realize that participation in the self-insurance program cannot be continued if the plant is being operated in such a manner as to jeopardize the self-insurance plan.

The self-insurance program, if properly administered with adequate provisions for proper policing of the various locations, can provide a direct means of control for fire safety purposes. Because it is management-oriented, the program may be more successful than insurance engineering programs developed for highly protected risks.

Although some self-insurance programs have undoubtedly been responsible for reduced losses, especially within the industrial community, there does not appear to be a rapid movement toward self-insurance on the part of either industry or government. However, once a self-insurance program is initiated, it appears likely to continue. Many organizations have operated on this basis for a number of years and have no intention of changing to a commercial insurance program.

Controls needed to ensure continued satisfactory operation include plan review and inspection functions and detailed analysis of all special hazards within the industry or institution. It is also necessary to have a soundly developed program for training maintenance personnel and others directly engaged in the day-to-day operation of the structures in question. Certainly, everyone connected with the location must be aware of the importance of compliance with fire safety standards and the insurance implications of any fire that may occur.

Educational Institutions

Several private universities offer programs related to fire safety. The Worcester Polytechnic Institute, with its Center for Firesafety Studies in Worcester, Massachusetts, has carried out research in a number of fire safety matters and offers a master's and a doctorate in fire protection engineering.

Fire Department Associations

Although made up primarily of members of public agencies, regional, state, and local fire associations are actually private organizations. These associations exist in practically every state and province.

Many of these associations have both career and volunteer fire department membership. In some states, career and volunteer fire service groups have organized separate associations functioning within the same state. In either case, fire prevention may well be considered

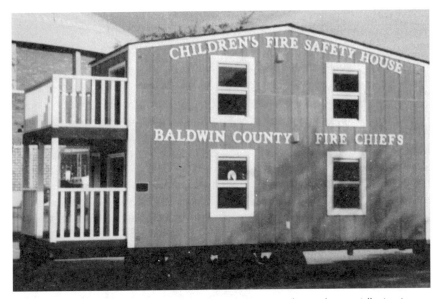

Figure 9-1 The local home builders association was the major contributor to the construction of this fire safety house in Alabama.

a major activity of the organization. A number of states have fire chiefs associations, as well.

Generally, a fire prevention committee is one of the standing committees of the association. Its function is to encourage the promotion of fire prevention by the member departments. The committee may hold meetings for the purpose of distributing fire safety information, and in some states it conducts an annual fire prevention contest. Member fire departments are urged to send in entries descriptive of the fire prevention activities carried out in their jurisdictions.

An example of a fire prevention group within a state fire association is the Fire Prevention Committee of the Maryland State Firemen's Association. This committee carries out an active program of fire prevention within the state and conducts several fire prevention contests each year. These contests include fire prevention activities by individuals as well as the fire companies within the state. Other committee activities also contribute to the fire prevention emphasis by the fire departments of the state.

The South Florida Fire Safety Association is another example of a public-oriented private organization having a bearing on fire safety in the state. The organization's members represent both public and private sectors, primarily in the southernmost counties of Florida. The Association, which was formed in 1972, holds monthly meetings as well as annual 40-hour seminars on fire safety subjects.

A number of county fire associations likewise sponsor active fire prevention programs. These programs may directly reach the public or may merely be designed to encourage fire prevention activities on the part of member companies.

Home Building Industry

The home building industry has assisted in the promotion of improved fire safety measures. An example is Castle Pines Property Management, Inc., in Castle Rock, Colorado. This developer has built a private and self-sustaining community south of Denver. The homes in this community are provided with automatic sprinkler systems as a matter of procedure at the time of construction. All builders are required to provide such systems in commercial structures as well as in private dwellings. This requirement is imposed by the developer in the interest of fire safety. Home builders associations have supported the construction of a number of mobile fire safety houses.

Summary

Fire prevention efforts in the United States have been motivated predominantly by private organizations, rather than by public agencies, as in most other countries.

Among organizations with a primary function in fire prevention, the unquestioned leader is the National Fire Protection Association, which has membership in over 70 countries. It is a nonprofit educational and technical association devoted to promoting scientific methods of fire protection and prevention and circulating information on these subjects to guard against loss of life and property by fire. Among the many publications of the NFPA, probably the best known is the *Fire Protection Handbook,* which has become the bible of the fire prevention field. Others include educational material and various periodicals, records of fires and fire loss statistics, as well as over 250 publications on technical standards, codes, recommended practices, and model ordinances, which are developed, prepared, and undated under the guidance of more than 200 technical committees.

Among other fire prevention–oriented organizations that have contributed greatly to the cause are the International Association of Fire Chiefs, the International Association of Fire Fighters, the International Association of Arson Investigators, the Society of Fire Protection Engineers, and the National Safety Council.

Organizations with allied interests in fire prevention are noncommercial organizations such as the Chamber of Commerce of the United States; professional associations and institutes in special fields that prepare standards and procedures having some bearing on fire protection; and a long list of trade associations devoted to the promotion of trade in products and services that relate in some manner to fire prevention.

Building code development groups have made a major impact on fire prevention. Operating at the national level are the Building Officials and Code Administrators, Inc., the International Conference of Building Officials, and the Southern Building Code Congress.

The insurance industry has been a dominant influence in the development and encouragement of fire prevention measures in the United States, and there is no doubt that many improvements have come about as a result of their direct relationship to insurance rate structures. Here, too, the associations and mutual systems representing a number of insurance companies have made significant contributions.

Insurance Services Offices establish insurance rates and are responsible for the grading of municipal fire defenses. In all states, insurance rates are subject to review by the state insurance commissioner.

Underwriters Laboratories Inc. has long been an important force in fire prevention through its comprehensive testing programs for equipment, materials, and assemblies that relate to fire protection and safety and has earned an international reputation for the UL label.

Some private business organizations maintain extensive internal fire prevention programs for their own protection, and some elect to operate under a self-insurance program and assume risks of fire losses instead of depending on commercial insurance coverage. The federal government might be considered a leading self-insurer. A self-insurance program puts direct responsibility on management to be sure that all proper precautions are taken to avoid fire.

And last, the contributions of educational institutions that conduct courses in fire science and administration, and of fire fighters' associations that are active in fire prevention education, must also be counted in the monumental effort that has been made by the private sector in this country toward the cause of fire prevention.

CHAPTER 10

Fire Prevention Responsibilities of the Public Sector

In the United States and Canada, governmental agencies have played a less prominent role in fire prevention than have their counterparts in other countries. However, the trend toward heavily urbanized areas and economic changes has made greater governmental responsibility in the field of fire prevention necessary, and the influence of governmental agencies at all levels is steadily increasing.

In this chapter the role of the federal government as well as those of state, provincial, and local governments will be outlined. Rapid changes are taking place in the sphere of governmental responsibilities, and activities by government that were unheard of 25 years ago are commonplace today.

The major change has taken place at the federal level. Not too many years ago, federal responsibility was confined primarily to direct protection of federal properties and fire safety enforcement relating to a rather narrow group of activities involving interstate commerce. Amplifications of interpretations of the term *interstate commerce* have broadened the scope of federal responsibility and operation in that field. The findings and recommendations of the National Commission on Fire Prevention and Control, established by an act of Congress in 1968, are also having a strong influence on fire prevention responsibility in the public sector.[1]

The Advisory Commission on Intergovernmental Relations, which was created by Congress to monitor the operation of the federal system and to recommend improvements, studied the federal role in local fire protection. In its 1980 report, the commission noted several "policy streams" through which federal involvement in fire functions was increasing: protection of national forests, defense (especially civil defense), and science.[2] It further noted that consumer protection and rural economic development were used as vehicles for advancing fire programs; however, most activity has been carried out through the first three streams.

Internal fire prevention efforts are broad in scope and encompass practically all federal agencies in some manner. A number of larger federal government installations maintain their own fire departments with personnel having responsibility for fire safety inspections, enforcement of fire prevention regulations, and public fire prevention education, in addition to their fire suppression duties.

Standards for fire safety within the federal government in the United States are generally set by individual agencies. Agencies maintain guidelines and regulations that must be adhered to by their employees. Regulations and standards of construction vary within the federal government, although the advent of the General Services Administration as a property management arm of government has probably brought about more uniformity.

Many federal agencies include fire prevention and evacuation training as a part of the indoctrination program for new employees. The responsibilities of certain federal agencies, especially the armed services, demand a high degree of attention to fire prevention because of the nature of the equipment they employ. The seagoing services, the U.S. Navy and U.S. Coast Guard, place a great deal of emphasis on fire prevention because of the grave problems related to fires aboard ship. The same condition prevails with aircraft assigned to the military services.

Although primarily aimed at enhancing fire prevention within federal facilities, government efforts in internal fire prevention have had a profound effect on fire prevention behavior of the public. There is no question that fire safety training and actual fire experiences while on active military duty have made a lasting impression on large segments of the present civilian population.

In a similar way, military reservists and National Guard members are exposed to fire prevention education through weekly or monthly drills, which may include an occasional fire prevention lecture. Likewise, training programs offered to federal employees can be of value in home fire prevention and fire response conditioning.

Federal agencies with fire prevention responsibilities that have an impact on members of the general public include practically all cabinet departments, as well as a number of independent agencies. Following are summaries of the fire prevention work being done by the various governmental departments and agencies.

U.S. Department of State

The Department of State has had several foreign aid programs relating to fire prevention as part of public safety efforts. These programs, which have operated under the Agency for International Development, have resulted in the imparting of recommendations and funds for the building of fire suppression and prevention forces in certain under-privileged areas of the world.

The Fire Protection Section of the State Department's Office of Foreign Buildings has responsibilities for ensuring that proper fire safety precautions are taken in embassies and legations throughout the world. In an effort to better accomplish its mission, the Section has developed a hand-held inspection system known as WANDA. By utilizing a scanner wand, the inspector can swipe an appropriate bar code, which represents a recommendation for inclusion in the final report. At the end of the inspection the inspector can transmit the day's activities to a report generator that creates a finished and formal report. This is a sample of innovations developed by federal fire inspection agencies. The system is now in use in some local fire and building departments.

U.S. Department of the Treasury _____

The Department of the Treasury, through three of its agencies, engages in a number of activities related to fire prevention that have an effect on members of the general public.

Customs Service

The Customs Service, which is under the Department of the Treasury, maintains control over merchandise, including fireworks, imported from other countries. The Service maintains representatives in all ports of entry. Their agents determine that markings for flammability, for example, appear on imported textile products as required by the federal labeling provisions. They are also active in the prevention of smuggling.

Bureau of Alcohol, Tobacco, and Firearms

The Bureau of Alcohol, Tobacco, and Firearms of the Department of the Treasury enforces federal statutes relating to explosives. The agency regularly inspects and licenses explosives dealers, manufacturers, and users when any aspect of the operation may involve interstate commerce. Display fireworks, for example, which are purchased in one state and transported to another, require possession of a license from the Bureau of Alcohol, Tobacco, and Firearms. Personnel of this bureau are empowered to investigate certain cases of arson, explosions, use of fire bombs, and careless activities relating to explosives. This function often brings these agents in contact with local and state fire department and law enforcement personnel.

Internal Revenue Service

Responsibilities of the Internal Revenue Service include the determination, assessment, and collection of internal revenue taxes. In performing these duties, agents of the agency may uncover attempts at fraud that included set fires. A number of fraud by arson cases have been thwarted by this agency.

U.S. Department of Defense

The Department of Defense includes the Army, Navy, Air Force, and Marine Corps, which is part of the Navy. These elements are considered to be military departments within the Department of Defense. In addition, several other agencies are within the Department.

Department of the Army

An Army activity having a bearing on fire prevention and protection is the maintenance of explosives ordnance disposal units stationed throughout the country. These highly trained units are available on request of local fire or police personnel to remove and render harmless military explosive devices that may be found in a community.

Several research projects carried out by the Army have likewise contributed to the field of fire prevention. These include the development of fire-retardant materials.

Department of the Navy

The Department of the Navy, which includes the Navy and Marine Corps, has made a substantial contribution to the fire prevention field through extensive research activities. While much of the research relates to the development of fire suppression devices and equipment, some effort has been directed toward fire prevention. Fire suppression research is, in fact, an effort to bring about more effective control of fire, which contributes to fire prevention through reduction of fire spread.

Department of the Air Force

The Department of the Air Force has devoted a great deal of research to the development of specialized equipment for fire suppression. Funding and technology have been available to permit extensive testing of new approaches in fire suppression before they are put to widespread use.

As with the Navy, a number of Air Force fire suppression advancements have had a residual effect in fire prevention. The very nature of fire suppression devices and procedures has the indirect effect of calling attention to the subject of fire.

U.S. Department of Justice

An agency of the Department of Justice, the Federal Bureau of Investigation pursues activities that contribute to the cause of fire prevention.

Federal Bureau of Investigation

The Federal Bureau of Investigation, an arm of the Department of Justice, has for many years provided laboratory examination services to local law enforcement agencies. Examinations conducted include those of materials found in and about a fire scene. Many individuals have been prosecuted for arson and unlawful burning based on examination of evidence by the Federal Bureau of Investigation Laboratory.

The Federal Bureau of Investigation has been of considerable assistance in arson investigation through the services of the National Crime Information Center and the extensive fingerprint identification program carried out by the agency. With authority over fugitive felons crossing state lines, the bureau has been of assistance in apprehending wanted arsonists.

U.S. Department of the Interior

The Department of the Interior maintains two bureaus whose functions directly and indirectly contribute to public safety and fire prevention education.

National Park Service

The National Park Service maintains a program to encourage conservation and environmental awareness in the national parks. This program, offered to the public, includes fire prevention as an environmental consideration. The exposure of the public to fire prevention measures and programs in the national parks undoubtedly has a residual effect outside of the parks.

Bureau of Land Management

A good portion of the nation's federally owned lands in the western states are administered by the Bureau of Land Management. This agency provides fire prevention and fire suppression support in connection with lands it manages.

U.S. Department of Agriculture

The federal department that has traditionally had the greatest fire prevention impact on the general public is probably the Department of Agriculture. The publicity campaigns of several of its agencies—the Forest Service, Soil Conservation Service, and Extension Service— reach directly to most citizens.

Forest Service

The U.S. Forest Service of the Department of Agriculture maintains a widespread fire suppression force and conducts an extensive nationwide fire prevention promotional program. This program, which has made Smokey Bear a national figure, reaches a high percentage of the U.S. population through television, radio, and newspaper advertising, as well as through signs and billboards.

Smokey observed his 50th birthday in 1994. His message, "Only YOU can prevent forest fires," is recognized by 95 percent of adults and 85 percent of children aged 8 to 12. The number of accidental wildfires caused by humans has been cut in half since Smokey arrived, despite the fact that the number of visitors to forests has grown tenfold during this period. The original Smokey was a cub who survived a New Mexico wildfire.[3]

The Forest Service fire prevention program has been widely accepted by the public, as demonstrated by a number of polls. State forest fire personnel have cooperated in the federal efforts and have on many occasions promoted forest fire prevention before school groups and civic organizations. The impact of the federal Forest Service program is more completely covered in Chapter 12.

Soil Conservation and Extension Services

Closely related to fire protection is the program of the Soil Conservation Service of the Department of Agriculture. This agency has been responsible for establishing countless farm ponds, which are put to use for fire protection by rural fire departments. Extension service programs include emphasis on fire prevention. The National 4-H program, for example, includes farm fire prevention themes.

U.S. Department of Commerce _____

The Department of Commerce had the major role among federal agencies in the fire safety field when the National Fire Prevention and Control Administration was part of it. In 1979 this agency was transferred to the newly established Federal Emergency Management Agency. The Department continues to have an important role in fire safety, however.

Bureau of the Census

A major function of the Bureau of the Census, an arm of the Department of Commerce, is to take a census of the U.S. population every 10 years. Data collected may serve a variety of needs in the fire prevention field. Varieties of fire experiences within different types of communities and income levels and comparisons of effectiveness based on population densities are among the possible uses of the census records.

National Institute of Standards and Technology

The National Institute of Standards and Technology has played an increasing role in fire prevention as a result of the enactment by Congress of the Fire Research and Safety Act of 1968.

As an arm of the Department of Commerce, the Institute, formerly known as the National Bureau of Standards, has a long history of cooperation with state and local governments in establishing standards of performance. As an example, the program of weights and measures, which greatly affects the marketing of all items sold on a basis of weight or other measurement, has long been under their purview. Although specific weight and measurement control programs are enforced at the state or local level, the overall standardization among states is based on procedures established by the National Institute of Standards and Technology.

The Institute's Building and Fire Research Laboratory is responsible for carrying out a great deal of research in the fire safety field. The Federal Fire Prevention and Control Act of 1974 strengthened programs previously operated by the Bureau in this field.

Programs of the Building and Fire Research Laboratory are carried out by five divisions: Structures Division, Building Materials Division, Building Environment Division, Fire Safety Engineering Division, and the Fire Science Division. The goal of the research of the fire-related divisions is to provide technical and scientific knowledge on all aspects of fire to reduce human and economic costs related to unwanted fires.[4]

U.S. Department of Labor _____

Under the Occupational Safety and Health Act of 1970, the Department of Labor embarked on an enforcement program that brings it in close contact with a high percentage of places of employment within the United States. This act probably has more direct impact on fire prevention enforcement at the local level than any federal legislation previously enacted.

Occupational Safety and Health Administration

The regulations of the Department of Labor's Occupational Safety and Health Administration (OSHA) are applicable to all employees, businesses, industries, institutions, farms, and other places of employment that engage in interstate commerce. The term *interstate commerce* includes all places that, for example, obtain supplies from, or market to, businesses in other states. For practical purposes there are few places of employment not covered by this broad definition. OSHA is charged with developing occupational safety and health standards and regulations. OSHA has promulgated a comprehensive set of regulations, including a number that incorporate National Fire Protection Association standards. Flammable liquids, liquefied petroleum gases, building exits, and electrical safety are among the fire protection subjects covered.

The legislation that established OSHA also established a procedure by which state governments can, if adequately prepared, enforce the provisions of the act within their jurisdiction. A number of states are utilizing this arrangement. In those that are not, the Department of Labor provides direct enforcement with their own personnel. Federal matching funds assist states that provide their own enforcement personnel. Local fire prevention inspections, as well as those of state agencies, are not preempted by this federal program, according to procedures set forth by the agency.

Mine Safety and Health Administration

The Mine Safety and Health Administration has a direct responsibility for fire prevention and safety within coal, metallic, and nonmetallic mines, often sharing this responsibility with a state bureau of mines. Records show that fires and explosions in coal mines have resulted in some major tragic losses of life. Inspection and fire prevention procedures in mines follow basic principles of fire prevention, but special consideration must be given to the unusual structural and atmospheric conditions inherent in mining operations.

U.S. Department of Health and Human Services _____

The Department of Health and Human Services has had a major impact on fire safety. This department, which was formed in 1953, has a number of agencies with direct interests in fire prevention.

Health Care Financing Administration

Responsibilities for ensuring the provision of proper fire safeguards in health care facilities receiving funds through Medicare (Title 18) and Medicaid (Title 19) rest with the Health Care Financing Administration. Fire prevention standards have been established for facilities housing patients in either program. Thousands of patients in long-term care facilities and hospitals are funded under these programs.

State agencies assist the Department of Health and Human Services in ensuring that fire safety requirements are met in these facilities. Funding may be discontinued for lack of compliance with the requirements, which encompass the Life Safety Code of the National Fire Protection Association.

U.S. Department of Housing and Urban Development _____

One of the responsibilities of the Department of Housing and Urban Development is the development and direction of programs to bring about lower-cost housing. The department has devoted considerable attention to fire protection for lower-income housing units and has funded several research projects in the fire protection field. Requirements for manufactured housing are also enforced by this department.

Federal Housing Administration

The Federal Housing program, an arm of the Department of Housing and Urban Development, has set certain standards for structures insured under the National Housing Act and other federal laws. These include minimum fire safety standards. Funding is provided by the agency for the construction of housing for the elderly, nursing homes, intermediate care facilities, and nonprofit hospitals. Minimum fire protection standards are included in these mortgage insurance programs.

U.S. Department of Transportation _____

The Department of Transportation, formed in 1966, includes a number of agencies that have an interest in fire prevention. Several of these agencies operated independently before the Department of Transportation was established.

Coast Guard

The U.S. Coast Guard has long had a legal responsibility in the field of fire prevention. Under the port security program of the Coast Guard, regulations pertaining to fire prevention and other security features are enforced in the ports of the United States. The Coast Guard has a responsibility for the inspection and regulatory enforcement on all waterfront facilities. This function means that a Coast Guard inspector has concurrent jurisdiction with municipal and state fire prevention agencies. The regulations enforced include requirements for fire extinguishing equipment, marking of extinguishers, maintenance of access for fire fighting equipment, control of cutting and welding operations, and control of smoking.

The Coast Guard also has a fire safety function in connection with vessels operating in U.S. waters. Vessels constructed in American shipyards are under the constant inspection of the Coast Guard during construction. A major portion of the obligatory inspection relates to fire safety. Once the vessel is in commission, the Coast Guard continues to inspect and ensure proper maintenance of fire fighting equipment and training of crews.

The loading and storage of cargo, especially that of a hazardous nature, is also under the purview of the Coast Guard. Coast Guard personnel may be detailed to directly supervise the loading and stowage aboard ship of explosives and other dangerous commodities. Comprehensive regulations relating to the transportation and the storage of dangerous articles of all descriptions are enforced by the Coast Guard.

Under the provisions of the Federal Boating Act of 1958, Coast Guard boating safety teams are required to inspect small boats to ensure compliance with required safety measures, including fire safety. The Coast Guard also has responsibilities in the field of abatement of oil pollution and other environmental issues.

In addition to surveillance over vessels, the Coast Guard is responsible for ensuring that merchant marine personnel have qualified through training. Licenses are issued to such personnel for the performance of specific duties after satisfactory completion of a comprehensive examination. These examinations, both written and practical, include a number of fire safety problems.

The Coast Guard also conducts investigations into causes of marine disasters, including fires. Appropriate sanctions are imposed where indicated.

Federal Aviation Administration

The Federal Aviation Administration is responsible for the control of all aspects of aviation, including safety features incorporated within aircraft. The agency ensures, for example, that carpeting in the cabin of passenger aircraft does not have an excessive flame spread. The agency is also concerned with the provision of adequate fire rescue forces and equipment at commercial aviation facilities.

Federal Highway Administration

The Federal Highway Administration, through its motor carrier and highway safety sections, controls the safe transportation of dangerous cargoes, including explosives and flammables, in interstate commerce over highways. This agency conducts inspections to ensure compliance with regulations covering safeguards for such movements. All commercial carriers employing motor vehicles come under this program when their operations extend to interstate commerce. The motor carrier and highway safety sections are responsible for the investigation of accidents involving controlled carriers.

Federal Railroad Administration

The Federal Railroad Administration is responsible for the regulation of railroads. The Bureau of Railroad Safety has specific responsibility for enforcing regulations on safety and fire prevention on interstate railroads.

Research and Special Programs Administration

The Research and Special Programs Administration (RSPA) coordinates Department of Transportation responsibilities pertaining to pipeline safety and the transportation of hazardous materials. Responsibilities include the administration of safety regulations, enforcement, and issuance of exemptions and interpretations.

RSPA's Office of Pipeline Safety establishes and provides for enforcement of safety standards for the transportation of hazardous and gaseous materials by pipeline. In the pipeline safety program, states may be awarded up to 50 percent of their costs in carrying out enforcement and inspection programs.

RSPA is the Department of Transportation's focal point for regulations, exemptions, research, coordination with the states, and the emergency response information published in its guidebook. Cooperation with state agencies is especially important if the state itself incorporates Department of Transportation regulations within its motor vehicles code, which a number of states do.

The *Emergency Response Guidebook* is the most widely distributed technical guide for initial response actions in the event of incidents (spill, explosion, fire) involving hazardous materials. This publication, which is revised triannually, is distributed to emergency responders through a network of key state agencies.

U.S. Department of Veterans Affairs

The Department of Veterans Affairs although not directly reaching the public at large with respect to fire prevention, has performed work that is of value in the field. Extensive research by the Veterans Administration in the use of noncombustible bed clothing is looked on as a progressive step, and fire prevention criteria developed by the Department are used as guidelines by many nongovernmental hospitals.

Independent U.S. Government Agencies

The Consumer Product Safety Commission has endeavored to eliminate hazards associated with consumer products, including toys, hazardous substances, and flammable fabrics. The agency has an enforcement and educational program that strives to reduce injuries resulting from use of common consumer products, most of which are transported through interstate commerce. The agency enforces regulations requiring proper labeling of hazardous substances and is authorized to legally ban from the market commodities that pose severe fire or safety hazards. It enforces the Flammable Fabrics Act as well. This law prohibits interstate marketing of wearing apparel and other materials that do not conform to flammability standards.

The Consumer Product Safety Commission has been active in the field of fireworks control and has banned the manufacturing and distribution of the more explosive fireworks. Flammable liquids for household use have also received the scrutiny of the Commission. Enforcement officers for the agency seek out items having unusual characteristics from a safety standpoint. The activities of this agency are having an increasing impact on fire safety.

The Environmental Protection Agency has the responsibility of ensuring the protection of the environment through the abatement and control of pollution. The comprehensive charge to this agency, organized in 1970, includes development of standards for open-air burning and a number of other fire-related pollution problems.

The Federal Communications Commission, although not as directly involved in fire prevention as some other agencies, has served the cause of fire prevention by its mandatory requirements for public service time on radio and television broadcasting stations. Many fire departments have taken advantage of this provision and have prepared public education fire prevention programs for television and radio. The agency also has contact with the fire service through control of radio frequencies used by public safety agencies.

The Federal Emergency Management Agency is the federal government's focal point for emergencies. In addition to the U.S. Fire Administration, which includes the National Fire Academy, the agency is responsible for emergency management and other coordinating programs relating to major disasters.

The Fire Research and Safety Act of 1968 provided for fire research activities and for the establishment of the National Commission on Fire Prevention and Control. The commission had the task of recommending to the President and to Congress procedures for reducing fire losses in the United States.

After extensive study, the commission submitted a report entitled *America Burning* to the president and Congress. This report, issued in 1973, contained as a major recommendation the establishment of "an entity in the Federal Government where the Nation's fire problem is viewed in its entirety, and which encourages attention to aspects of the problem which have been neglected."

Needs that the Commission felt could be addressed by such a federal agency included more emphasis on fire prevention; better training and education in the fire service; more education of the public about fire safety; and recognition of the hazardous environment to which Americans are exposed from a fire safety standpoint. They also addressed the need for improvements in building fire protection features and in research activities related to fire safety.

Congress subsequently enacted legislation in 1974 establishing the National Fire Prevention and Control Administration as an agency of the Department of Commerce. Under the provisions of President Carter's 1979 reorganization plan, the National Fire Prevention and Control Administration was transferred basically intact to the new Federal Emergency Management Agency. The National Fire Prevention and Control Administration became the U.S. Fire Administration under separate legislation.

The National Fire Academy, which is housed with the Emergency Management Institute at the National Emergency Training Center in Emmitsburg, Maryland, is responsible for advanced-level fire training activities. A wide variety of programs is offered on campus and in field classes offered throughout the country. This academy has gained international stature and is recognized as having major impact on fire service activities in the United States. Both volunteer and career fire service personnel can take advantage of course offerings. A number of programs relate to public fire education, code enforcement, arson suppression, and fire prevention administration.

The U.S. Fire Administration has also played a major role in improving fire safety efforts. The Administration has led efforts to increase the popularity of residential sprinkler systems and has assisted state and local agencies in pilot public fire education programs. The latter program is known as the Community Volunteer Fire Prevention Program and emphasizes the involvement of private citizens rather than public safety personnel.

The Fire Administration also operates the National Fire Incident Reporting System. This system, which is described in more detail in a subsequent chapter, has enabled fire safety personnel to study the causes of fires and to develop more effective public education programs.

Juvenile fire setters have been the focus of another program of the Fire Administration. Fire service personnel have been trained in methods aimed at reducing such fires through effective counseling and education.

From time to time the Fire Administration has conducted special studies of major fires and has looked at specific fire problems in high-risk areas. Efforts have been made to disseminate information on these matters to bring about improved community fire safety.

The Federal Trade Commission administers regulations pertaining to advertising that have been helpful in eliminating false descriptions of products that may be dangerous or that do not perform properly. In several cases, the commission has acted against the manufacturers of small throwaway fire extinguishers to which labels indicating superior performance had been attached.

The General Services Administration serves as a management agency for the construction and operation of government-used buildings, the procurement of supplies, and the control and disposal of records. In carrying out these responsibilities, fire protection standards have been established by the Public Buildings Service, an arm of the General Services Administration.

The National Aeronautics and Space Administration has made valuable contributions to fire prevention and public safety in connection with space research. Highly reliable fire-retardant materials developed for use in spacecraft have been adapted for commercial purposes and are available in a number of products on the market. As the fire-retardant qualities of these materials become better known, there should be a wide public demand for their use in everyday products.

Figure 10-1 Federal, state, and local agencies often cooperate in the investigation of major fires. (Ocala, Florida, Fire Department.)

The National Institute of Building Sciences was established by Congress in 1978 for the purpose of providing "an authoritative nationally recognized institution—to advise both the public and private sectors of the economy with respect to the use of building science and technology.[5] As a quasi-governmental agency, the Institute depends on private as well as federal support. Board members are selected by both groups, the president appointing 6 of the 21 members and the remaining positions being selected by the private sector.

The Institute has been quite successful in getting groups with diverse interests to work together in the common causes of improved building safety and more economic construction costs. It also advises other federal agencies on matters relating to building construction. Fire safety has been a high priority of the Institute.

The Nuclear Regulatory Commission regulates and licenses the civilian use of nuclear material and generally has a responsibility in connection with the safety of this material. Fire prevention procedures are a factor in these regulations.

The American Red Cross, another quasi-official agency, has joined in national fire prevention campaigns and actively supports programs for the reduction of fire losses.

The National Transportation Safety Board has specific statutory responsibility for the investigation of accidents in civil aviation as well as highway, rail, marine, and pipeline accidents. Investigation is mandatory in aviation accidents; other accident investigations are made on the initiative of the board.

Members of the board, who are appointed by the president, have executive authority in conducting their investigations and also have the authority to conduct formal proceedings for review of appeal on the suspension, amendment, modification, revoking, or denial of any certificate or license issued by the secretary of transportation or by an administrator operating with the department. The Safety Board's reports of their investigations of major accidents have resulted in changes in regulations and in changes in procedures within the Department of Transportation.

Canadian Government Agencies _____

In comparison with U.S. federal fire programs, Canada's Fire Commissioner coordinates the work of the provincial fire marshals and fire commissioners without direct administrative oversight. The agency, a part of Human Resources Development Canada, is housed in Ottawa and is responsible for enforcing fire protection requirements at certain federally owned facilities.

In addition to having fire protection coordination and enforcement duties, the Fire Commissioner compiles an annual report of fire losses in Canada. This publication has long been looked on as a leader among national fire loss compilations.

Canada has one military service known as the Canadian Forces, which includes military aviation, seagoing, and land elements. Fire safety is under the direction of the Canadian Forces Fire Marshal. All fire protection and prevention activities within the military service are under this command.

In Canada, fire research activities are carried out by the National Fire Laboratory of the Institute for Research in Construction, National Research Council Canada. Their program deals with fire behavior, characteristics of materials in fire, and fire resistance of structural elements in buildings. They have also conducted research in other areas related to fire safety.

The Canadian government regulates transportation safety through Transport Canada. This organization has an active program for protecting the public from hazards of commodities being conveyed by rail, truck, air, and ship. The Canadian Coast Guard is involved in fire safety control for vessels built and registered in Canada.

Explosives are regulated at the national level in Canada through the Explosives Branch of Natural Resources Canada. All regulation of explosives relating to manufacture and transportation is handled at this level.

The Product Safety Branch of Health and Welfare Canada is responsible for fire safety in products manufactured for consumer use. This includes household, personal sports, and recreational uses. One of the major issues addressed by this agency has been that of tent fire safety. It has also been involved in research in furniture and textile safety as well as in a number of other areas usually associated with consumer protection.

State Agencies

All state and provincial governments have agencies with responsibilities in the field of fire prevention. Code enforcement and inspections at the state level have been responsible for a multitude of improvements in fire safety measures and fire prevention practices.

State Fire Marshal

The state agency normally having fire prevention as a major responsibility is the state fire marshal's office. Forty-nine states now have an organization carrying out fire safety functions at the state level; such an agency does not exist in Hawaii. All provinces and territories in Canada have such offices.

State fire marshals' offices were first established shortly before the turn of the century. Since that time they have grown in number at a rather steady rate, with offices established in Missouri and Arizona in 1972. Some states have established fire marshals' offices as a direct result of some major tragedy; others have done so because of a realization that effective fire prevention must be carried out as a function of the state government.

In Arkansas, Michigan, Oregon, and Pennsylvania, the state fire marshal's office is an arm of the state police. In three of these states, the agencies carry out a full range of fire marshal's duties, including fire prevention code enforcement and fire investigation. In Pennsylvania, responsibilities in the inspection and code enforcement field are limited to flammable liquid installations; other agencies are assigned the remaining inspection responsibilities.

Because of the past close association of fire prevention with the insurance industry, the office of the fire marshal in a number of states is assigned to the office of the insurance commissioner. In some states—Florida, for example—the insurance commissioner serves as ex officio state fire marshal; in others, such as Georgia, the fire marshal is appointed by the insurance commissioner. Another reason for this close relation with the insurance department is that in some states the office of the insurance commissioner and fire marshal is funded by a tax on insurance premiums.

In an increasing number of states, the office of the fire marshal has become a part of the state department of public safety, an arrangement that makes possible closer cooperation with other state law enforcement agencies, especially in the field of fire investigation.

Indiana has combined the offices of state fire marshal and state building commissioner into a Department of Fire Prevention and Building Safety. This action came as a result of conflicts that had arisen in plan review functions and is looked on as a model method of achieving coordinated statewide fire and building safety delivery.

In addition to duties normally carried out by state building and fire safety agencies, the department also has the responsibility for boiler and pressure vessel enforcement, fire service training, elevator safety, and industrialized building regulation.

The state fire marshal and building commissioner exist as offices within the department. Plan reviews are coordinated, as are public building and fire safety education functions. An executive director oversees the department and supports the lawfully assigned duties of the state fire marshal and state building commissioner. The department also has a 16-member Fire Prevention and Building Safety Commission, which is appointed by the governor. Membership includes representatives of the fire services as well as construction and design fields. The commission adopts rules and regulations and hears appeals from enforcement actions.

Texas has established a state Commission on Fire Protection, which oversees the State Fire Marshal, public fire safety education, fire service personnel qualifications, funding distribution to fire departments, the fire reporting system, and inspection of communities to determine insurance rates. Commission members represent career and volunteer fire services and nongovernmental fire safety interests. Fire service training and forestry fire protection are arms of Texas A&M University.

Although the location of the fire marshal or commissioner's office in the government may vary from Canadian province to province, their responsibilities include code enforcement, investigation, fire reporting system operation, and fire service training in all provinces and territories.

Personnel Appointments. Considerable diversity exists in the ways fire marshals or the individuals in charge of the fire marshal's office are selected. In some states, the position is filled by gubernatorial appointment; in others, appointment is made by the state fire board or fire prevention commission, or by the attorney general, state comptroller, or other person under whom the fire marshal's office operates. Length of tenure varies from an indefinite period to four, five, or six years. In states in which the fire marshal's office is an arm of the state police, the appointment is usually based on departmental assignment procedures. Alabama's fire marshal is a merit system position.

Appointment methods for deputies and other personnel within the fire marshal's office are also varied. In some states, all deputies are merit system or civil service employees; in others, the deputies are appointed to serve at the pleasure of the fire marshal. In the latter situation, suggestions on appointments may come from someone at a higher level in the state government, with possibilities of political involvement in the appointments. This was fairly common practice in past years. Now, however, in a majority of the states, public employees are selected under a merit system or other nonpatronage employment program.

Selection by civil service or merit system has a twofold advantage: the state is generally assured the services of qualified personnel, and the employee benefits from continuous employment regardless of the political party or individual in control of the administration at the moment. This is not to say that political appointment always produces bad results. Loyalty to the administration may produce diligence in performance during the tenure of office. But even though an appointee may be well qualified in background and ability, regardless of political considerations, there is always the problem of interruption of service caused by termination of employment in a political changeover in administration.

Fire Marshal's Responsibilities. Although the responsibilities of the office of the fire marshal vary from state to state, its primary function is usually to enforce state fire prevention codes and to investigate suspicious fires. The office is also normally the coordinating agency for all fire prevention activities within the state government. The state fire marshal strives to obtain compliance with fire prevention regulations in connection with the licensing program of other state agencies, such as the health department and department of education. The fire marshal may recommend the inclusion of fire safety requirements in licensing and accreditation procedures and in other state controls imposed on facilities within the state.

Fire Code Enforcement. In a number of states, the office of the state fire marshal is the primary fire code enforcement agency. In some states, a locally designated fire marshal or fire chief may serve as an ex officio deputy to the state fire marshal for the purpose of code enforcement within the local jurisdiction.

Fire code enforcement at the state level may on occasion have advantages over strictly local enforcement. Local political pressures may be less effective in hampering fire prevention code enforcement administered at the state level than at the local level. The state fire marshal's office can serve as a backup for the local fire marshals' offices, to render assistance and support in difficult situations. The state fire marshal's office can also lend support in the form of technical advice and services to smaller communities, whose budgets do not usually allow for specialized personnel.

Fire Investigation. The state fire marshal's office has the responsibility in most states to conduct investigations of all fires to which it is called. Notification procedures vary; however, local fire and police departments are usually the source of calls.

Under the Model Fire Marshal Law,[6] which is enacted in many states, the state fire marshal or a representative of the office has right of entry for the investigation of fires and responsibility for control and suppression of arson within the state. A number of states invest the fire marshal with powers of subpoena and arrest; under these legal conditions, the office is considered to be a law enforcement agency within the framework of state government. The state fire marshal's office may provide investigators for investigation of suspicious fires. The agents are usually qualified as expert witnesses in the courts and can play a major part in the successful prosecution of a case. In many states, fire marshal's office personnel are considered peace officers and are provided with sidearms. These individuals have the power of arrest in arson cases.

Legal Aspects of Code Enforcement. The office of attorney general is responsible in most states for providing legal assistance in enforcing fire prevention regulations and codes. In a few states, this responsibility includes direct prosecution of arson cases. In several states, the fire marshal's office is an arm of the attorney general's office. Code enforcement cannot be effective without legal backing and advice. The attorney general's office may play a major part in bringing about fair and reasonable code enforcement practices.

Fire Statistics. Another duty carried out by most fire marshals' offices is the compilation of fire statistics. An effort is being made to encourage every state to engage in fire recording and to develop statewide statistical data for use in fire safety programs.

Several states have statutes requiring fire departments to record all fires within the state; in others, such an obligation is also placed on insurance carriers doing business in the state. Certainly, it is necessary to have reports both from fire departments and insurance carriers to develop an accurate statistical compilation of fire losses and insurance data. A review of reports submitted by insurance companies will reveal many losses that were not reported to the fire department.

Fire Code Adoption. In most states a state fire prevention code in effect. These codes are promulgated by the state fire marshal, the insurance commissioner, the state fire board, or some other agency granted the power to promulgate regulations. Usually, a public hearing is required so that the public, special interest groups, and fire service personnel have the opportunity to address the fire marshal or fire prevention commission regarding possible problems or advantages they see in connection with implementation of the requirements.

Promulgating regulations by commission or board is much better in many aspects than initiating requirements through statutory provisions. Statutory requirements are far less flexible and may be misleading to individuals who attempt to interpret the limited amount of wording in the statute.

The fire marshal's office may depend heavily on the regulations of other state agencies. For example, department of health regulations may include specific fire requirements that are actually enforced by the state fire prevention bureau or fire marshal's office. The same procedure may be used with the fire safety provisions of the motor vehicles code and of other government agencies.

Fire Safety Education. A number of state fire marshals operate public fire education programs. In some cases state efforts are directed toward the development of local capabilities in this field, while in others, fire marshal's office personnel provide programs directly to the public. Ohio's fire marshal has long been a leader in this field.

Other Duties. A number of state fire marshal's offices have been given the responsibility for enforcement of manufactured and mobile home regulations as well as those pertaining to industrialized housing. Georgia and Tennessee have comprehensive programs in this field. Although these controls relate to fire safety features, plumbing, sanitation, structural stability, and other phases of construction are also included. This means that the fire marshal's office must have an ability to handle these related areas as well. The responsibility for inspecting all electrical work is assigned to the fire marshal in some states including Tennessee and Wyoming.

Explosives control activities are assigned to the fire marshal in some states. This control may include licensing programs for people possessing and using explosives, as well as a capability to deactivate clandestine devices, bombs, and other explosive devices.

The advent of greater federal participation in fire safety has brought about an interest in improved coordination of fire safety services at the state level. Among concepts advanced is that of a state coordinating commission or council whose purpose is to influence fire safety efforts in the state. The council or commission may or may not have direct authority over any operating agencies, such as the fire marshal's office or state fire training program. Such a commission would likely have as representatives fire chiefs, both career and volunteer fire fighters, building officials associations, the state fire marshal and director of fire service training, the state forester, and members of other statewide agencies and organizations that have responsibilities and interests in fire protection.

Health Department

All 50 states have a department of health in the state government. The health agency has an interest in fire safety as a part of its responsibility in safeguarding citizens from health hazards and, more specifically, in ensuring proper safeguards for facilities under the direct control of the agency: hospitals, extended-care facilities, and other such institutions. The health department may also have an interest in fire prevention through its air-quality program, which in most states includes certain bans on open and outside burnings.

In view of the immobile nature of occupants, institutions under health department control have been required in practically all states to upgrade fire protection to ensure adequate safety for patients. The state health department is in a position to exert strong influence in the field of fire safety because of its many interests in the health field. The work of the sanitarian, for example, in checking the cleanliness of a restaurant has a bearing on fire safety. A clean restaurant is less likely to invite fire damage.

Education Department

Each of the 50 states has a department of education. This agency has an interest in ensuring the safety of all schools within the state, including those that are privately operated but come under state accreditation control. As a rule, state department of education regulations require periodic fire inspection and plan reviews for new construction, evacuation plans, fire drills, and other procedures to strengthen fire protection and fire prevention. In some states, school regulations are issued and enforced in cooperation with the office of the state fire marshal.

Departments of education may also be responsible for including fire prevention education in the required curriculum. Education departments frequently permit school visits by fire department personnel and encourage personnel to prepare children for lifelong awareness of the importance of fire prevention.

Department of Labor and Industry

All state governments have a division for regulation of labor and industry. Although primarily responsible for industrial safety, these agencies often include fire prevention in their enforcement and educational programs. With the advent of the Occupational Safety and Health

Act of 1970, the responsibilities of the department of labor and industry have increased in many states. In a number of states, this agency has been designated by the governor as the primary enforcement and coordinating agency for state enforcement of the provisions of OSHA.

In several states—Pennsylvania and Wisconsin, for example—in which the fire marshal's duties do not include a full range of fire code enforcement, inspections relating to fire prevention are the responsibility of the department of labor and industry. Likewise, the control of explosives may be in this agency.

Industrial safety classes organized and sponsored by departments of labor and industry may include emphasis on fire prevention. North Carolina has a comprehensive industrial safety education program that stresses fire prevention as an element of safety.

A number of states have enacted "right to know" laws. Under these laws employers are required to furnish their employees information on toxic substances in the workplace and are required to provide training on the safe handling of such substances in emergency procedures. Employers must notify local fire departments of the location and characteristics of all toxic substances regularly present in the workplace. This legislation may affect fire prevention bureau activities in that much of the information relates to that sought during code enforcement visits.

Federal legislation likewise requires that information relating to toxic materials be given to employees and to the local fire services under the Superfund Act of 1987. Fire safety efforts are enhanced by greater distribution of information relating to hazardous materials.

Public Utilities Commission

The public service or public utilities commission is another state agency that has an interest in fire prevention. These commissions normally regulate public utilities, including gas and electrical distribution. Both of these utilities have a bearing on fire safety within a state. The public service commission in some states regulates transportation of hazardous commodities. Again, a function having some bearing on fire prevention is exercised by the commission.

In several states the public service commission has a statutory responsibility for conducting electrical inspections within the state or for seeing that such inspections are conducted properly. Electrical inspections are most essential in any comprehensive fire prevention program.

Racing Commission

Although not found in every state, the racing commission is another agency that may have some influence in the field of fire prevention. Requirements of fire safety may be imposed by the commission as a condition of continued state recognition of racetracks. An interesting feature of code enforcement is raised in this phase of activity. To ensure continued operation of a track, steps must be taken to provide a greater degree of life safety for horses than would be required for animals under the life safety code in effect in the state. Such enforcement creates a dual standard in that animals owned by a farmer are not required to be protected in the same manner as horses used for racing.

Bureau of Mines

States having operating mines generally have a state agency responsible for control of the industry. The bureau of mines usually includes fire prevention as a part of its legally constituted duties. Fire prevention and protection in underground mines is a field quite remote from structural fire prevention. State bureaus of mines have fire prevention programs designed primarily for personnel working underground.

Forestry Division

Forest fire prevention is a major function of divisions of forestry in all states and provinces. These agencies are generally responsible for fire prevention and control in their areas of jurisdiction. They are normally established in such a way as to permit assignment of personnel to fire prevention or suppression, depending on immediate demands. During fire seasons, for example, all personnel are assigned to fire suppression duties, whereas when fires are less frequent, these employees may be used to visit schools, campsites, and public areas for fire prevention education purposes.

Forestry personnel may be assigned to a state agency that also has a responsibility for the operation of parks. Park operation provides yet another avenue for the implementation of fire prevention programs in that visitors to the parks can be encouraged and, in fact, required to adhere to fire safety practices. Fire prevention efforts of forestry departments in the 50 states have undoubtedly been responsible for the dissemination of fire safety information to countless individuals who would not otherwise have been contacted in any manner by personnel having such a message.

Other State Agencies with Related Functions

A number of state agencies have functions that are less directly related to fire prevention. A rather obscure example is the agency that examines and licenses barbers. By including fire prevention as a requirement for licensure, the agency can bring about improved fire prevention practices in this type of occupancy. Although not a high life-hazard occupancy, the effectiveness of state backing for fire code enforcement can be helpful to local fire prevention officers.

The functions of several other state agencies in this category are described in the following paragraphs:

State Medical Examiner. Although perhaps not thought of in this regard, the state medical examiner or postmortem examiner actually performs a function having a bearing on fire prevention. The physicians assigned to such an agency are responsible for performing autopsies to determine the cause of death and factors relating thereto.

All fire deaths should be investigated, and an autopsy should be performed. Through this means, a factual picture of our fire prevention problems as related to life safety can be more readily obtained. In one state, for example, where such a program was vigorously implemented, it was learned that a number of deaths from causes other than fire were reported as fire deaths. Individuals in some cases had been killed by weapons before the setting of the fire. Had autopsies not been performed, the deaths would have been carried as fire deaths. However, a body damaged by fire is not pleasant to examine, and in states where nonprofessional coroners are used, there may be a tendency to automatically write the death off as "smoking in bed" or some other fire-related cause when in fact the victim died from a cause unrelated to the fire. Another responsibility is that of determining whether the victim died from carbon monoxide poisoning before being burned in the fire. Again, many deaths each year are reported as "deaths from burns" that occurred as a result of carbon monoxide poisoning, with subsequent burns after death. An autopsy is the only way to avoid such an error.

Commissions on Aging. Most states have an agency responsible for problems of concern to the elderly. These agencies can be of great assistance in ensuring the dissemination of fire prevention information to elderly citizens. Thousands of elderly people who are not living in institutions are contacted on a regular basis by these agencies.

State Institutions. State agencies with responsibilities for housing patients, inmates, and other wards of the state are likewise responsible for fire prevention measures in connection with the operation of their facilities. Through the cooperation of inspecting agencies, such as the fire marshal's office, fire safety practices in these institutions can be maintained at a high level.

Although organized as a part of the health department in some states, the agency responsible for mental health must be mentioned.

This agency usually operates a number of institutions housing mental patients, and certain steps are taken to achieve fire safety within these facilities. Mental health agencies often have responsibility for licensing privately owned structures housing people suffering from various mental disorders. The provision of fire safety requirements will have a bearing on the safe operation of the facilities.

Gaming Commission. A number of states have gaming commissions, which regulate casinos, riverboats, and other gambling operations. Usually, these agencies require strict adherence to fire safety requirements in gaming facilities. As an example, when South Dakota made gambling legal in Deadwood, the state fire marshal was given the responsibility for requiring many fire safety improvements in existing buildings.

Building Code Agencies. A growing number of states are enacting a state building code on either a mandatory or permissive basis, with functional responsibility in a variety of state agencies. Under the provisions of some state code, a local jurisdiction may, if it deems it desirable, adopt the state building code for enforcement within the jurisdiction.

An example of a building code enforced on a mandatory basis at the state level is the code of Wisconsin. Although mandatory at the state level, local enforcement is permitted.

Educational Institutions. The college and university system operated by state governments has a very definite role in fire prevention in two kinds of programs: those conducted for the safety and well-being of students residing on campuses and occupying classroom spaces; and fire service training programs to improve the public fire departments.

As an example of a program in the first category, the University of Illinois at Urbana maintains a full-time, well-equipped fire department for the protection of the campus. Personnel assigned to this department are also responsible for conducting fire safety inspections on the campus. The department is also active in fire prevention education.

An example of the second type of program is that carried out by Texas A&M University at College Station. In addition to providing training to the members of the public fire service in Texas, the Texas A&M fire service training program places major emphasis on training industrial personnel in fire prevention and fire suppression. Special programs for industrial and institutional groups are offered on a staggered basis, and certain programs are offered in Spanish for the benefit of Latin American fire fighters.

State fire training programs operated by colleges and universities are among the most effective in the fire service training field. By being housed on campuses, they have the advantage of the tie-in with academic programs and of offering participation in research work that might not otherwise be available. Foremost among these fire service training programs are those conducted by the University of Maryland, Iowa State University, and Oklahoma State University, as well as Texas A&M University, mentioned before. Several colleges and universities offer four-year degree programs in fire protection.

In some states, fire service training programs are operated as arms of the state department of vocational education. This arrangement is particularly prevalent in states having programs in which federal vocational education funds are used in connection with fire service training activities. Community colleges also offer fire safety–related programs.

Judicial Branch. The judicial branch of state government has a bearing on fire prevention through legal processes in fire prevention code enforcement and arson control. Legal action in code enforcement is effective only with the backing and support of the judicial branch of government.

In view of the wide variety of organizational concepts in the judicial branch from state to state, no attempt will be made to attach court titles to responsibilities; however, fire code enforcement is at the lower-level court in most states because of the common designation of such violations as misdemeanors. Arson cases are normally heard by a middle-level court—generally the highest-level court within the county—because of the classification of arson as a felony.

Legislative Branch. The legislative branch of state government must be mentioned because of the major role it plays in enacting legislation relating to fire prevention, fire and arson control, and its role in enacting budgets for fire prevention and fire protection. As an example of the latter role, in one state the legislature has over several years appropriated amounts in excess of $0.5 million per year for the purpose of providing fire protection and fire prevention measures in the state's mental institutions.

County Agencies

Fire safety activities at the county level are quite varied from one county to another.

An increasing number of county governments are operating fire departments for the protection of the entire county, or at least for portions outside of incorporated cities. The Los Angeles County Fire Department protects unincorporated areas as well as a number of incorporated cities with which it contracts to provide fire protection. Fire prevention services are also provided.

Although some metropolitan counties provide a full range of services and activities similar to those normally associated with municipal governments, the majority of the counties in the United States are not heavily involved in fire prevention programs.

Many counties provide fiscal support for volunteer fire department operations and some maintain fire marshals' offices, communications centers, and training programs to assist local fire departments.

Because so-called full service counties are rather rare, the programs they offer may be considered with the discussion of municipal government activities, which follows. An ever-expanding number of counties are enforcing planning and zoning requirements at the county level.

Municipal Government _____

Municipal government has a tremendous impact on fire prevention and fire control. There are few municipalities, regardless of area or population, that do not have some provisions for public fire protection in the form of an organized fire department, which may have career, volunteer, or both types of personnel. The municipal fire department is without question the major bulwark against the perils and ravages of fire in the United States and Canada today.

Today even the most remote communities have fire departments. It is rare to find an incorporated community with more than 1,000 people that is not served by such an organization. A fire department in a community is not always dependent on expenditures by the municipal government. In a number of states the majority of fire departments are established and funded entirely apart from the municipal government. In these places the fire department is often incorporated, owning its property and electing its own officers. Public funds may provide support for these departments, but in many cases no public funds are used.

The role for the fire department in fire prevention generally is the same whether the department is tax-supported or not.

Other municipal agencies having an interest in fire prevention include the police department, whose involvement may be similar to that of the state law enforcement agencies previously mentioned. A city's department of public works has a close relationship with fire safety. The department of public works is usually responsible for the maintenance of streets. Waterworks and water distribution systems, whether publicly or privately operated, are major factors in public fire protection.

A smaller municipal agency having some interest in fire prevention is the bureau of weights and measures, which may also be found at the county and/or state level. Weights and measures have a bearing on fire prevention because of the importance of accuracy in operations dispensing flammable liquids and gas. An overfill can result in spillage of a flammable product, with the possibility of fire ensuing.

The planning and zoning agency is usually structured at the municipal level and plays a major part in the fire protection of a city or town. Zoning regulations are predicated on public safety concepts and cover the maintenance of proper clearances between structures, hazardous industrial processes, storage of flammable liquids, etc. Enforcement of zoning regulations can do a great deal to prevent the existence of a conflagration hazard within a community.

The municipal building department is usually responsible for examination of plans and inspection of construction, a function that has major impact in fire prevention because the building codes relate so directly to construction from a fire safety standpoint.

Electrical inspection is a municipal function almost entirely devoted to fire prevention and safety. The inspection of new and existing structures for compliance with the electrical code is a major factor in the reduction of the number of fires caused by improperly installed wiring.

Plumbing inspection at the local level also has a bearing on fire prevention in that fire protection requirements can be adequately met only through properly installed piping. This agency may be responsible for inspection of all piping in a city.

As with the state-level operation, the municipal health department may have a function in the fire prevention field. A sanitarian in the health department is responsible for eliminating trash and debris in public places, thereby reducing fuel for maliciously and carelessly set fires.

Many municipalities operate public utilities such as electrical, telephone, steam, and transportation service in addition to utilities normally associated with municipal government, such as sewer and water services. These public agencies may be used to disseminate fire prevention literature and to render other assistance in the fire prevention field. Transportation agencies, for example, can assist in fire prevention by posting fire prevention signs on transit vehicles.

Support Agencies

Certain agencies operating at all levels of government have what may be called an indirect association with fire prevention. Government organizations responsible for taxation and assessment and for fiscal operation of government at all levels provide important support for effective fire prevention. Their part in ensuring funding of necessary programs cannot be taken lightly.

Agencies such as emergency management, art commissions, transportation, beautification groups, local and state environmental agencies, historical preservation boards, and a number of others at some time have some involvement with fire prevention and should not be overlooked for community cooperation.

Summary

Government, except at the municipal level, has not been as active in fire prevention and protection in the United States and Canada as it has been in other countries. However, with population growth and economic changes in the 20th century bringing about greater concentration of the population in urban areas, the influence of governmental agencies in fire prevention efforts at all levels is steadily increasing. Indicative of this trend is the creation by Congress in 1974 of the National Fire Prevention and Control Administration.

Among federal agencies, the Department of Agriculture has traditionally had the greatest impact on the general public with respect to fire prevention (mainly through the Forest Service). However, the National Institute of Standards and Technology and the Federal Emergency Management Agency have made many contributions in the field.

Under the Occupational Safety and Health Act of 1970, the Department of Labor has impact on fire prevention through enforcement procedures of the act, which applies to all employees, businesses, industries, institutions, farms, and other places of employment that engage in interstate commerce.

The Department of Transportation is also an important force in fire prevention. The Department of Defense has maintained fire prevention efforts and training for members of the armed forces, which carry over into civilian life.

Among state agencies, the office of the state fire marshal is paramount, having wide responsibilities in fire code enforcement and inspections, and investigations of suspicious fires, among other duties. The state fire marshal also lends aid and support to municipal fire departments.

At the municipal level, the community fire department is without question the major bulwark against the perils and ravages of fire in the United States and Canada today. Practically every community has such a force, whether it is paid, volunteer, or a combination of the two.

Notes

1. U.S. National Commission on Fire Prevention and Control Report, *America Burning* (Washingtion, D.C.: U.S. Government Printing Office, 1973).
2. Advisory Commission on Intergovernmental Relations, *The Federal Role in Local Fire Protection* (Washington, D.C.: U.S. Government Printing Office, 1980).
3. "Smokey Bear Turns 50," *Parade* (February 6, 1994).
4. National Institute of Standards and Technology, *NIST Building and Fire Research Laboratory* (Gaithersburg, Md., 1993) p. 4.
5. National Institute of Building Sciences, *Annual Report to the President of the United States* (Washington, D.C.).
6. *Model Fire Marshal Law* (Quincy, Mass.: Fire Marshals Association of North America). This document gives a comprehensive description of the recommended duties and powers of the state fire marshal.

CHAPTER 11

Fire Prevention through Arson Suppression

Suppression of arson is clearly a factor in fire prevention efforts. In many communities arson is a major part of the fire problem; therefore, any steps that can be taken to prevent arson reflect favorably on the community's fire statistics.

Although law enforcement agencies often share with fire services the responsibility for suppressing arson, the latter service is usually the prime mover in arson control. Not only can arson be discouraged by effective investigation and prosecution, but many communities have mounted successful programs aimed at preventing such incidents. In this chapter various aspects of arson control are discussed; however, this is not intended as a complete treatise on the subject.

The Crime of Arson

As a type of criminal activity, the maliciously set fire historically has been considered a serious crime. As an example, the first arson law in Maryland, enacted by the General Assembly meeting in St. Mary's City in 1638, classified arson as a felony. Punishment was death by hanging, loss of hand, or being burned on the hand or forehead with a hot iron. The offender also forfeited ownership of all properties. If the offender escaped death for the first offense, there was a mandatory death sentence for the second offense.[1]

In Illinois, which obtained statehood in 1818, arson was considered so serious an offense that between 1819 and 1827 it was one of only four crimes punishable by death.[2]

Early western towns were likewise hostile toward arsonists. Vigilante justice occasionally superseded the established system. Closely built frame buildings, coupled with meager fire suppression capabilities, could spell doom for the entire town when an arsonist was at work.

A suspected arson ring in Virginia City, Nevada, was dealt with by a vigilante band that stormed the local jail in 1871 and removed and lynched a suspected arsonist. Other members of the arson ring were sentenced to the state prison. Arson did not occur frequently after that time.[3]

Arson is one of the few criminal activities in which there is usually no immediate recognition that a crime has been committed. A fire is usually considered to be an accident until a thorough investigation shows evidence of arson. The civil disorder fire is a different matter. In this type of maliciously set fire, it is readily apparent in most cases that a crime has been committed. But the investigator must still prepare a tight care that examines as far as possible ways in which the fire could have been started accidentally.

The crime of arson has another unusual feature. The weapon most often used is readily available. This weapon is the ordinary match, which is, of course, available at most every convenience store, service station, and restaurant. No program could be developed to control the acquisition of matches, nor would such a cause have popular support.

Arson Statistics

The *Uniform Crime Report,* published by the Federal Bureau of Investigation of the U.S. Department of Justice, provides interesting statistics on arson in the United States. Statistics for 1992 are used in this chapter; however, a study of statistics for other recent years will indicate that there is a fairly constant pattern in arson statistics for the past five or six years.

A total of 102,009 arson offenses were reported to the Uniform Crime Reporting Program in 1992. Arson is defined by the Program as any willful or malicious burning or attempt to burn, with or without intent to defraud, a dwelling house, public building, motor vehicle or aircraft, personal property of another, etc.

Only fires determined through investigation to have been willfully or maliciously set are classified as arsons. Fires of suspicious or unknown origins are excluded.

The arson rates ranged from 103 per 100,000 inhabitants in cities with populations over 1 million to 20 per 100,000 rural county inhabitants. The suburban counties and all cities collectively recorded rates of 36 and 56 per 100,000 inhabitants, respectively. Overall, the 1992 national arson rate was 48 per 100,000 population.

Regionally, the highest arson rate was registered in the western states, with 60 offenses per 100,000 population. Following were the northeastern states, with a rate of 47 per 100,000; the midwestern states, with 45 per 100,000; and the southern states, with 41 per 100,000.[4]

The increasing arson problem is clearly shown in a review of the statistics regarding incendiary and suspicious U.S. building fire losses as published by the National Fire Protection Association. The figures indicate that there were 5,600 incendiary or suspicious building fires in 1950, with a loss of $15 million. By 1975 arson fires had soared to 144,000, with a $533.9 million loss and a loss of life of over 1,000.[5]

By 1984, U.S. arson losses had risen to $1.417 billion; however, fatalities as a result of arson had dropped to 530. In 1992 $2 billion in arson property losses were reported, as well as 605 fatalities.[6]

Wildland fire statistics likewise include a high percentage of arson fires. For example, arson is the cause of almost a third of the forest fires in Florida each year.

Model Arson Laws

Arson is best discussed in terms of the laws under which the crime is investigated. The model arson law as developed by the Fire Marshals Association of North America is a part of the laws of many states.[7] This law is divided into four sections as to degree.

First-degree arson under the model law is the burning of a dwelling. Under this category, any person who willfully and maliciously sets fire to, burns or causes to be burned, or aids, counsels, or procures the burning of a dwelling is guilty of arson. The model law implies that the offense is the same whether the building is occupied, unoccupied, or vacant. It further indicates that the section applies whether the property is the property of the accused or of some other person. The section includes structures that are parts of the dwelling, such as kitchens or shops. The sentence provided under this section is generally from 1 or 2 years to 20 years.

Second-degree arson under the model law is burning buildings other than dwellings. This section also requires that the fire has been willfully and maliciously set, and it provides that an individual is chargeable who has either set the fire, caused it to be burned, or has aided, counseled, or procured the burning of the building. Again, the offense is applicable whether the property belongs to the person charged or to another. The statute often provides for a possible penalty of 1 to 10 years.

Third-degree arson in the model law pertains to the burning of other property. Individuals who willfully and maliciously set fire to, burn or cause to be burned, or aid, counsel, or procure the burning of any property of any class, as long as the property has a value of $25 or more and is the property of another, are guilty and may be sentenced for a period of not less than one nor more than three years in the penitentiary.

Fourth-degree arson in the model law relates to attempts to burn. This provides that a willfully and maliciously set fire may be chargeable as an attempt even if combustible and/or flammable materials have been only distributed. This includes any acts preliminary to the setting of the fire. The suggested law also includes definitions to clarify the recognition of such a crime.

A separate offense category is included for the crime of burning with intent to defraud an insurer. This likewise requires that the act was willful and malicious and that the property be insured at the time of the fire.

It should be noted that the term *willfully and maliciously* occurs time after time in the model arson law. This means that a fire set accidentally is not chargeable under the model arson law. To obtain a conviction of arson, a prosecuting attorney must be able to prove in court that the act was willfully and maliciously perpetrated. In fact, the law enforcement officer should not consider making an arrest on an arson charge if such willful and malicious action cannot be proved. A distinction is made in the law between dwelling places and other occupancies. There is a chance of death in the case of a fire that has been maliciously set in a dwelling.

A number of states have adopted the Model Penal Code of the American Law Institute as first published in 1962.[8] This code covers the entire criminal law field and encompasses some concepts that are at variance with more traditional modes.

In the Model Penal Code, arson is covered in the following manner, as excerpted from Arkansas Criminal Code:

5-38-301.—ARSON

(a) A person commits arson if he starts a fire or causes an explosion with the purpose of destroying or otherwise damaging:

(1) An occupiable structure or motor vehicle that is the property of another person; or

(2) Any property, whether his own or that of another person, for the purpose of collecting any insurance therefor; or

(3) Any property, whether his own or that of another person, if the act thereby negligently creates a risk of death or serious physical injury to another person; or

(4) A vital public facility.

(b) Arson is a Class B felony if any property sustains less than twenty thousand dollars ($20,000) worth of damage; a Class A felony if any property sustains no less than twenty thousand dollars ($20,000) nor more than one hundred thousand dollars ($100,000) worth of damage; and a Class Y felony if any property sustains damage in excess of one hundred thousand dollars ($100,000).

(c) For purposes of this section, "motor vehicle" means every self-propelled device in, upon, or by which any person or property is, or may be, transported or drawn upon a street or highway.

5-38-302.—RECKLESS BURNING

(a) A person commits the offense of reckless burning if he purposely starts a fire or causes an explosion, whether on his own property or that of another, and thereby recklessly:

(1) Creates a substantial risk of death or serious physical injury to any person; or

(2) Destroys or causes substantial damage to an occupiable structure or another person; or

(3) Destroys or causes substantial damage to a vital public facility,

(b) Reckless burning is a Class D felony.

5-38-303.—FAILURE TO CONTROL OR REPORT DANGEROUS FIRE

(a) A person commits the offense of failure to control or report a dangerous fire if he knows that a fire is unattended and is endangering the life, physical safety, or a substantial amount of property of another person, and he:

(1) Fails to act in a reasonable manner to put out or control the fire when he can do so without substantial risk to himself; or

(2) Fails to act in a reasonable manner to report the fire.

(b) Failure to control or report a dangerous fire is a Class B misdemeanor.[9]

Most statutes patterned after the Model Penal Code use similar language. A comparison of the Model Arson Law and the arson sections of the Model Penal Code will reveal that there are advantages and disadvantages in each.

Arkansas and all other states also have statutes relating to wildland fires. In some states a wildland arson is considered a felony, while in others it is a misdemeanor.

Arson laws in Canada, like all criminal laws, are uniform throughout the country. Such uniformity exists in most other countries of the world.[10]

Motives for Arson

Although not absolutely necessary, it is most helpful to the prosecution of a case if a motive can be established for the setting of fire. A primary motive is avarice, a great desire for wealth. In such a situation, the perpetrator feels that a settlement from the insurance company can be made on the loss without the insurer realizing that the insured has in fact set the fire to defraud the insurer. This type of arson occurs most frequently during times of poor economic conditions; however, there are exceptions to this rule.

Another motive is revenge. An individual sets a fire in a building to get back at the owner or a building occupant. Another reason for a

Figure 11-1 Arson motives as shown in the annual report of a major city. (Annual Report, Detroit, Michigan, Fire Department.)

revenge fire being set is a dispute involving lovers. In cases on record, the victim's body has been saturated with flammables and ignited, with death ensuing.

Closely related motives are spite and vanity. In the latter case, the fire setter attempts to prove his or her own importance.

Psychological maladjustment is often found to be the cause of incendiarism. The fire setter may receive gratification from setting the fire or from feeling superior as the sole possessor of information regarding the origin of the fire. There is also a possibility of a fire being set for the purpose of attracting fire apparatus with red lights and sirens operating. The fire setter may receive gratification from seeing the red lights or from hearing the sirens.

Some individuals actually enjoy seeing the fire burn. Such individuals usually stay in the background while the fire is being fought; however, they usually observe everything and enjoy the flicker of the flames and the smoke from the fire. Such individuals may be called pyromaniacs when they can no longer resist the urge to set fires.

Some fires are set to conceal other crimes. Homicide, burglary, breaking and entering, and larceny are among the most frequent crimes for which a fire is set as a cover. These fires are often difficult to investigate, with a good possibility of the crime being successfully covered.

Thousands of fires are set merely for "kicks." Vacant buildings often are targets. Perpetrators are motivated primarily by a desire for a little excitement in an otherwise routine existence. This is primarily a motive of juveniles and young adults. It differs from other motives in that several individuals set the same fire. These are often called *vandalism* fires.

Fires set by individuals for political reasons are discussed in detail later in this chapter. This type of activity is not difficult to recognize.

Investigation of Suspected Arson

Successful investigation of maliciously set fires requires the establishment of a corpus delicti. This means that the investigator must be able to prove that a crime has in fact been committed. Every potential natural cause for such a fire must be eliminated so far as possible. It will be necessary to prove beyond a reasonable doubt that the fire could not have occurred as a result of accidental or natural causes. In defending the accused, the defense attorney in an arson case will most often attempt to prove that the fire was accidental. The prosecuting attorney must make every effort to prove that the fire could not have started accidentally.

Responsibility for Investigations

Fire investigation responsibilities are generally carried out in one of three ways:

1. Entire job, including cause determination; investigation and arrest carried out by fire service investigators.
2. Entire job, including cause determination; investigation and arrest carried out by police service investigators.
3. Split responsibility, with fire service personnel being responsible for cause determination followed by police investigation and arrest. In some communities a team concept is employed; in others fire service personnel merely notify the police when arson is suspected.

Many major cities, including New York, Los Angeles, Denver, Miami, and Houston, assign complete responsibility for arson suppression to the fire service. Personnel assigned must be thoroughly trained in law enforcement procedures and are empowered to make arrests. This concept has an advantage because the entire responsibility is placed on one agency and it eliminates the shifting blame for poor performance and the jealousies that may arise when two agencies are investigating the same crime. Of course, communities utilizing this concept continue to rely on police support, primarily in intelligence and patrol activities.

In some communities, police departments are vested with the entire responsibility for arson detection and investigation. This system, to be successful, entails a great deal of training for police officers in examining fire scenes. The officers must qualify as expert witnesses in examining fire scenes to obtain any convictions. A person with fire suppression experience usually qualifies much more easily.

Responsibility assigned to a joint police-fire team works well in many communities. However, it has failed in others. A problem with this concept is that work schedules in police and fire departments are generally not the same. Days off are not the same, which results in lost time when one of the team members is not on duty. If one team member goes ahead with the investigation and successfully concludes it, the other may be unhappy.

Some communities have no team but merely an arrangement whereby the fire suppression officer or fire marshal, on determination of arson as the possible cause of a fire, contacts the police department. The police department assumes all responsibility for the subsequent investigation and possible arrest. Both agencies must testify in court for a successful prosecution. Press publicity often gives credit only to one of the agencies, causing resentment on the part of the unmentioned agency.

Smaller communities often depend on a state or county fire marshal's office for fire investigation services. Local fire officers or the state fire investigator may make the determination of arson. In either case the state investigator carries out the subsequent investigation and makes an arrest where appropriate. This is a desirable arrangement, especially in view of the difficulty for persons who have investigated very few fires to be qualified as experts in court. Some smaller municipalities have banded together and developed an "on-call" system for mutual aid in fire investigations in an effort to overcome the experience problem.

Mississippi law requires each of the state's 82 counties to have at least one deputy sheriff trained in arson detection and investigation. These individuals, who are trained by the state's fire academy, are able to conduct a preliminary investigation and summon a state fire investigator when appropriate.

Charlotte, North Carolina, established a unique fire investigation task force, which includes fire and police department investigators, agents from the U.S. Bureau of Alcohol, Tobacco and Firearms, and the North Carolina State Bureau of Investigation as well as a district attorney. The task force has maintained a case clearance rate above 30 percent since its formation in 1985.

State fire schools as well as the National Fire Academy provide training for fire investigators. In addition, the International Association of Arson Investigators holds instructional seminars at both the national and state levels. Similar seminars are held in Canada.

Judges and prosecuting attorneys also need indoctrination in the arson control area. Seminars for these individuals have proven to be quite successful in creating a greater awareness of the intricacies of successful prosecution of arson cases. One state raised its arson conviction rates from less than 1 percent to 50 percent in less than three years by conducting seminars and mock trials for judges and attorneys.

Insurance Fraud

It is desirable to obtain a full inventory of materials and contents of structures. This is especially important if insurers should attempt to prove that a building was overinsured. Adjusters and other people familiar with evaluation techniques can be helpful in this connection. The records of insurance carriers should be checked to determine if there has been any recent change in insurance coverage on the building. Such a change may be an indication of an attempt to defraud the insurance carrier. To prove fraud, a claim must have been filed by the insured.

Insurance companies are concerned about soaring losses as a result of arson. In an effort to combat rising losses, which are reflected in higher insurance costs for the property owner, insurance companies have intensified their arson suppression programs.

One major insurance industry project is the Property Insurance Loss Register. This industrywide program is designed to record pertinent data about fires from adjusters. Computers store names of insured tenants, owners, partners, corporate officers; types of occupancy; cause, time, and date of losses, as well as insurers and amounts of coverage. This information is available for use in investigations. The program is supported by all using companies of the Property Claims Services of the American Insurance Association.

Insurance carriers sponsor arson control seminars and assist in forming statewide arson suppression associations in many states. They deny liability on suspicious fires, which places the insured in the position of having to sue to recover alleged losses. Insurance carriers often employ their own investigators to assist in such cases. In many cases the courts are upholding the denial of liability. In some cases the matter does not reach court, because the insured may merely drop the claim against the company rather than enter into litigation.

Some insurance companies assist communities in training accelerant-detection canines. These animals have proven to be quite successful in locating points of origin. Companies also provide major funding for arson award programs, in which tips leading to arrests are rewarded.

In Canada, the Insurance Crime Prevention Bureau has a team of investigators who assist with investigations on request. This organization has been quite successful in abating fraud fires.

The Value of Confessions

Occasionally in the investigation of an arson case, investigators can obtain a confession to the crime. Such a confession is valid if properly obtained and can be of great assistance in the successful prosecution of the case. No threats may be used nor can any promises be made in taking a statement that turns out to be a confession. It is extremely important that the individual be forewarned of full rights as required by the U.S. Supreme Court's *Miranda* decision. This decision requires that interrogation should not begin until the suspect has agreed to submit to the procedure. The individual must be warned, according to the decision, that a statement need not be made at all. The suspect should also be told that he or she may stop at any time and request the presence of an attorney; that in the event the suspect cannot afford a lawyer, one will be provided; and that any statement made could be adversely used in a court of law.

In the case of arson or malicious burning, a confession alone may be used for convicting the defendant if there is corroborative evidence connecting the individual to a fire. This means that it is not always necessary to have a witness who can place the defendant at the scene of the fire at the time of its burning. It may be sufficient to have a witness testify that the defendant was seen around the area 15 or 20 minutes before the fire or testify that the defendant was seen around the fire scene shortly after the discovery of the fire. In using such a witness, the prosecutor should consider the need for accurate times of sighting and other information placing the defendant at or near the scene within a short time before or after the fire.

Fire Scene Examination

Fire scene examination is a critical part of the investigation of fires. This examination cannot take place until the fire has been brought under control; however, the scene should not be left unguarded at any time. The fireground should be securely maintained until the investigation has been completed and no evidence or photographs at the scene are needed.

In examining the fire scene, every effort should be made to locate possible sources of ignition, such as electrical wiring, heating appliances, carelessly discarded cigarettes, or other sources that may have accidentally caused the fire. The investigator should be alert for any possible evidence pointing to the origin of the fire. Any suspicious items should be carefully preserved for laboratory examination and/or use in court proceedings. Proof must also be established that a fire did actually occur, and the investigator must have the owner or owner's agent testify as to the condition of the structure before the fire.

The fire fighter plays a major part in the investigation of a maliciously set fire. Without the complete cooperation of the fire department, successful investigation and prosecution will be extremely difficult. Fire suppression forces responding to the alarm often have the first trained eyes on the scene for observing conditions. The fire department may also observe conditions relating to the spread of the fire, the use of an accelerant, and other factors.

The conditions noted in the building during fire inspections and pre-fire planning surveys can be most helpful in determining any significant changes in conditions at the time of the incendiary fire. The fire inspector may have recorded the original position of cabinets and other equipment that may have been moved by the person starting the fire.

Fire company officers can observe the general conditions at the scene, such as weather and road conditions, on approach. These factors may have a bearing on the subsequent investigation (through consideration of tire tracks in the snow, as an example).

The first fire department personnel to arrive should observe people, automobiles, as well as license numbers and descriptions of vehicles in the vicinity. This is not an easy task for a fire fighter while fighting the fire or rescuing occupants of the building. But any remembered detail may be of great help to an investigation. Just one license number might be the right one to trace a suspected arsonist. And if an alarm is turned in promptly and response is immediate, a suspect can sometimes be arrested on the scene, especially because whoever started the fire is apt to stay around long enough to see that the job was successful.

Fire department personnel should also observe the color of the smoke and flames. This may contribute materially to the subsequent investigation of the fire. White smoke may indicate that hay, forestry products, or materials with vegetable contents are burning and may also indicate the presence of phosphorus. Yellow smoke may indicate the presence of film, sulfurous material, sulfuric acid, or smokeless powder. Black smoke normally indicates a petroleum product, including rubber or tar. Red flames will indicate a petroleum-base product, while blue flames usually indicate alcohol content or natural gas.

The first personnel on the scene should observe the size of the fire and the speed at which it is traveling. Again, as with noting license plate numbers, this is not easy to do under the pressure of fire control and rescue problems.

The fire officer should particularly observe whether separate fires may be burning, along with the intensity and rapidity of spread. This information may be extremely critical at a later time when investigation indicates that the fire was in fact maliciously set.

Other indicators at the scene of the fire that may be of value in an investigation are odors, such as those of gasoline or kerosene. Methods of extinguishing the fire might also have significance. The condition of windows and doors in the building should be observed—whether they are locked, indications of forcible entry into the building, or signs that material has been removed from the premises.

Bystanders at the fire scene may offer some clues. Familiar faces in the audience may have been noticed at other fires or emergencies in the community. The dress of the individuals on the fire scene is especially important if the fire is at an unusual hour; for example, between 2:00 and 3:00 A.M., bystanders who are fully dressed and wearing neckties would be suspect. They may have, of course, just happened to pass by at the time of the fire.

Fire Investigator

The fire suppression officer may, when the fire is under control, decide that a fire investigator should be called to the scene. This should be done before the fire suppression forces depart. To preserve evidence, full security should be maintained at the fire scene. Police officers, security guards, or other persons having a legal responsibility may need to stand by at the fire scene awaiting the arrival of the fire investigator. The scene should be guarded until the fire investigator has arrived.

The fire suppression officer, while awaiting the arrival of the investigator, may look for evidence of arson intent. Among things to look for are unusual odors in the building (possibly indicating the use of flammable accelerants), multiple fires, and undue charring and uneven burning. A trained investigator can readily note uneven burning and excessive or undue charring. The fire officer or investigator, or both, may look for holes that have burned through the walls or ceilings of the building. This would indicate the possible use of accelerants in the fire. There is a growing tendency for fire setters to deactivate sprinkler systems, block fire doors, and otherwise impede the operation of fire protection equipment in their efforts to make a successful job of arson. All of these should be checked to see if they have been tampered with. Equipment may be rendered inoperable by closing valves or by the use of plugs or other impediments to operation.

Tracks around the premises may be quite revealing. They can best be noticed by fire personnel arriving early at the fire. At least an observation may be made that there were people walking around the building before the fire. There is a good possibility that this will not be useful evidence, because the tracks can well be obliterated by fire service personnel carrying out fire suppression duties at the scene.

This is also be true of fingerprints. Although smoke and heat may obliterate most fingerprints, occasionally some can be found by thorough processing of the area. Most fire service personnel do not have training in fingerprint techniques and may overlook the possibilities of lifting prints.

A thorough and systematic examination should be made of the fire scene to determine possible causes of the fire and to detect evidence of tampering with any equipment. Heating equipment should be checked to see if gas valves were left open; electric and gas stoves should be checked for the possibility of burners having been turned on. Electrical equipment and appliances should also be examined.

Fatal fires deserve special attention. An autopsy should be conducted in all cases to ascertain the true cause of death.

A search should be made to detect any possible use of flammable liquids and for containers left behind at the fire scene that may have held flammable liquids. Ignition sources to look for are matches in the area, candles, or trailers placed between sources of ignition. Mechanical igniters or timing devices are not frequently found but have on occasion been used to start a fire. And not to be discounted is the possibility that the fire was started as a result of glass magnifying the rays of the sun. At the same time, excessive amounts of debris, rags, waste, and accumulations of papers or other combustibles in the area should be noted. All these factors should be carefully assessed and pointed out in the fire investigator's report.

Preservation of Evidence

Every effort must be made to preserve and protect any evidence of the cause of the fire that comes to light during the fire investigation. The evidence should, if at all possible, be maintained at its original location. It should then be removed, identified, and safeguarded by the fire investigator after arrival on the scene. Extreme care is needed in using water around the fire scene where evidence may be involved. This precaution is not always easy to observe. Care should be taken in salvage and overhaul operations. There is often a conflict between investigation and suppression activities. Overzealous cleanup activities sometimes result in the removal or destruction of necessary evidence. There should be a mandatory procedure for handling evidence so that it is not damaged or distorted. The owner or occupants of the building should under no conditions be permitted to return to it until the investigation has been completed. Visitors, including reporters, should not be allowed to enter the fire scene.

The use of barricades can be helpful if other means of restraining entry are ineffective. The fire service generally has the legal means of keeping unauthorized people from entering the structure while fire fighting operations, including overhaul, are underway.

The fire officer should make a note of the time evidence is discovered. If at all possible, some other person should also witness the discoveries and verify the conditions found. This verification may be a key factor in court.

The fire officer may also photograph the scene of the fire and have evidence gathered and placed in appropriate containers. It is normally more desirable for the fire investigator rather than the fire suppression officer to carry out this activity.

Control of Statements to the Press

Statements to the press regarding fire causes should be carefully controlled. Many investigations have been jeopardized because of the unwise release of information to the news media. Some fire service personnel have a tendency to make statements to the press regarding the details of fires without having specific knowledge of conditions. It is very easy to make comments to the press that later turn out to be incorrect. The most desirable approach is to have one individual responsible for all press releases. Statements about fire causes made by police officers assigned to crowd control and other authorities who are not members of the fire service may be most damaging to the prosecution should an arson case be developed.

Arson and Civil Unrest

Multiple arsons during periods of civil unrest during the 1960s and 1970s brought back an arson motive that had not been prevalent for many years. Such fires impose a severe strain on routine investigative procedures.

Tulsa, Oklahoma, and several other cities experienced disorders in the 1920s. At least 27 fatalities occurred in the 1921 racial disturbances in Tulsa. More than 1,000 buildings, primarily residences, were destroyed by fire. Detroit experienced riots during World War II.

In 1992, Los Angeles experienced the most expansive, violent, and costly epidemic of urban unrest in this century. A total of 53 civilians were killed and 2,328 were injured, including 59 fire fighters. Over 850 buildings were damaged or destroyed by fire, with a total loss of over $560 million. The Fire Sprinkler Association of Southern California estimated that nearly 2.5 million square feet of building floor area were saved by action of automatic sprinklers.[11]

In an effort to foresee possible fire problems, the fire investigator should participate in intelligence sessions. Intelligence-information gathering is not confined to disorders and civil unrest. Much can be learned about possible arson plans by organized crime, militant organizations, and other groups and individuals who may be bent on causing fires.

Operations during times of civil unrest or disorder require planning. Fire investigative personnel cannot merely sit and wait for calls and expect to have success once multiple fires start occurring. For example, it could be very useful to assign fire investigation personnel to patrol duties in the time of disorders. These patrols could detect fires early and thereby increase chances for successful suppression; they could also aid in apprehending those responsible for setting the fire. The use of unmarked vehicles is probably most desirable for this purpose. In a few cities, patrols of this kind have been equipped with portable fire extinguishers to enable them to control incipient fires. The possibility of such personnel being lured into an area for physical attack must be kept in mind in developing procedures.

Fires that occur during periods of disorder when feasible should be as promptly investigated as those that occur during normal times. A study of fires occurring during periods of unrest will usually show that many people are setting fires for the first time. It is easy to lay aside investigation of minor fires during a disorder in deference to fires having a great monetary loss or loss of life. Failure to investigate these smaller fires may only serve to improve the self-confidence of their setters and to allay their fear of being apprehended.

During periods of unrest there may be, in addition to a multitude of set fires, a tendency to turn in false alarms and bomb threats. These calls must receive attention, since it could be genuine. Individuals have not hesitated to use explosive devices during periods of disorder. False alarms may be intended to send fire equipment to the other side of town to permit the unabated spread of planned fires.

Fires must be investigated in such a manner as to provide adequate information in the event an arrest is made. The possibility of a property owner deliberately setting fire to the premises during a period of unrest cannot be overlooked. Fires occurring during civil disorder may encourage individuals who are naturally inclined to set fires. An impulsive fire setter, for example, could be strongly influenced by television publicity given to fires during disorders.

Fire investigation during a civil disturbance involves a number of unusual problems. Guarding the scene of the fire for evidence preservation may be extremely difficult. Angry mobs may make it impossible to maintain effective safeguards. Also, the same building may be set on fire several times during the course of the night. Procedures normally followed in preserving evidence must be laid aside. Members of the National Guard may be asked to guard the fire scene until the investigator can safely return.

Some communities have investigators respond to fire scenes along with or just behind fire apparatus during periods of disorder. This gives them an opportunity to locate and photograph evidence before it is removed. Fires set during disorders are usually rather crudely set.

Collecting and preserving evidence is just as important for fires set during times of civil disorder as for any other type of maliciously set fire. There is no real reason fires cannot start from natural or accidental acts during times of disorder as well as during less turbulent times. The case will probably come to trial under quieter conditions. The investigator must be prepared to testify as to whether the fire could have been of natural or accidental origin. At least, it must be shown that an examination of heating appliances, electrical systems, stoves, flammable-liquid operations, etc., did not indicate these to be the cause of the fire.

If flammable liquids were used, the investigator should determine, if possible, where the flammable liquids were purchased and the means used to transfer them to the premises. Containers are necessary to transport flammable liquids.

When many fires occur in a short period of time, a problem of handling of evidence generally develops. It is imperative that the evidence be adequately marked and properly stored. Materials gathered at the fire present a problem as portions of fire bombs may be quite fragile. A scoop-type box is useful for this purpose.

Arson Arrests

Many law enforcement and fire service investigators consider arson to be a difficult crime to solve with an arrest. While arson convictions may require diligent efforts, the *Uniform Crime Report* indicates that 15 percent of arsons were cleared by arrest in 1992. Southern states reported a 21 percent clearance rate; Midwestern, 10, Western, 13, and Northeastern, 12. Clearance rates compare favorably with the 13 percent rate for motor vehicle theft and 20 percent rate for larceny-thefts.

Success rates for arson clearances were greatest in rural counties (22 percent), followed by suburban counties (17 percent), and cities (14 percent). People under the age of 18 accounted for 43 percent of all arson clearances, an increase from 36 percent in 1985.

With respect to occupancies, the greatest percentage of arsons cleared by arrest was in community/public buildings (33 percent); while the lowest was in motor vehicles (7.8 percent). This clearly indicates the need for greater attention to motor vehicle arsons. In many jurisdictions there is little follow-through on these fires.

The *Uniform Crime Report* indicates that 16,322 people were arrested on arson and related malicious burning charges in 1992. In

1971, 11,154 were arrested on these charges. Forty-one percent were juveniles and were referred to juvenile authorities for adjudication.

Of those arrested on arson and malicious burning charges, 87 percent were males. This statistic is fairly constant from year to year. Among those arrested, 76 percent were white, 22 percent were black, and the remainder were of other races.[12]

Juvenile Fire Setters

As noted in the foregoing statistics, the juvenile fire setter is a major problem in the arson field today. The term *juvenile fire setter* includes both those children under the age of responsibility as well as those who can legally be held responsible for their acts. Some jurisdictions use the term "child fire setter" because the word *juvenile* may be associated with *juvenile delinquents,* causing parents to be reluctant to report their child's fire setting activities to authorities.

Studies have indicated that a high percentage of fires are set by juveniles, many by boys under the age of ten. A model program to address this problem is that organized by the Memphis, Tennessee, Fire Department. The program includes short- and long-term counseling, fire safety skill instruction, fire prevention projects, and parental awareness as well as extensive networking.

Figure 11-2 Fire service personnel counsel children regarding fire setting activities. (Gainesville, Florida, Fire Rescue Department.)

The Memphis program, known as "Fire Saver," incorporates the assistance of juvenile court, the board of education, child victim advocate program, fire department, Red Cross, Mental Health Institute, Medical Center Burn Unit, state Department of Human Services, Center for Abused Women and Children, prison reception center, and Charter Lakeside Hospital, a private facility that has provided major support to the network.

Children three and younger are counseled at the fire scene but may later be required to enter the program. Those over three are given juvenile summons and required to attend the Fire Saver program. Children who damage fire equipment in schools or turn in false alarms also must participate. Some children are found to have emotional disturbances, while others set fires out of mischief or as a criminal act.[13]

Court Decisions

Two U.S. Supreme Court decisions have a bearing on the investigation of fires. The 1978 decision, *Michigan v. Tyler,* established guidelines for right of entry to investigate cause and origin and to subsequently gather evidence should the fire be determined as incendiary.[14] The Court held that the original entry for fire suppression and subsequent investigation to determine cause and origin is permissible without a warrant; however, subsequent entries may necessitate obtaining consent to enter or a warrant.

In a later case, *Michigan v. Clifford,* the Court reaffirmed its findings in the earlier case and recognized the ability of the investigator to seize evidence in clear view at time of entry.[15] In that both cases directly relate to the investigation of fires, details of the decisions should be studied by fire investigators.

Following the lead of the U.S. Supreme Court, state courts have indicated similar restrictions. In a South Dakota Supreme Court case, *South Dakota v. Jorgensen,* the court upheld use of evidence found during the state fire investigator's initial entry while the fire was still smoldering but upheld suppression of evidence found in later, warrantless entries.[16]

Arson-Related Research Projects _____

The National Center for the Analysis of Violent Crime, located at the FBI Academy in Quantico, Virginia, conducted a comprehensive study of serial arsonists. *Serial arson* is defined as an offense committed by firesetters who set three or more fires with a significant cooling off period between the fires.

Researchers reviewed records of almost 1,000 incarcerated arsonists and in turn conducted detailed interviews of 83 confirmed serial arsonists in their places of confinement. The research was reported in *A Report of Essential Findings from a Study of Serial Arsonists.*

In addition to the sex and racial breakdown revealed by arrest records, the study indicated that almost half of the serial arsonists interviewed had some type of tattoo, while almost one-fourth had some form of physical disfigurement. Over 60 percent had multiple felony arrest records for offenses other than arson, and almost two-thirds had multiple misdemeanor arrest records. Over half had been held in juvenile detention in earlier years. One-fourth reported at least one suicide attempt.

A majority of the serial arsonists grew up in middle-class neighborhoods. Relationships with mothers were much closer than with fathers in a high percentage of the cases. Many started setting fires at very young ages. The 83 offenders interviewed had set a total of 2,611 fires, an average of 31.5 arsons per offender.

Of special interest to investigators is the fact that offenders reported they had been questioned by law enforcement officers an average of 3.7 times each before being arrested. The arsonists indicated that they had set an average of 25.3 fires without being questioned. They indicated that efforts by law enforcement investigators were responsible for 38 percent of the apprehensions. In 21 percent of the cases, the individual went to law enforcement agencies and confessed, while informants were responsible in 7 percent of the cases. Witnesses were the key element in 12 percent of the arrests.

Accomplices assisted in 20 percent of the fires. In a majority of cases (59 percent), material available at the scene was the accelerant of choice. Gasoline and other petroleum products were used in 28 percent of the fires. Matches were the predominant ignition device.

Thirty-one percent of the serial arsonists remained at the scene of the fire, while 28 percent went to another location. Slightly over half returned to the scene at some time after the fire, usually within 24 hours of the arson.

Nearly half of the serial arsonists consumed alcohol before setting fires. Over half cited revenge as the motive for their arsons. The next most prevalent motive was that of "excitement." Nearly one-half entertained no thought of ever being arrested for their crimes. Almost one-half admitted that their acts were premeditated as opposed to impulsive.

A review of the findings indicates that serial arsonists generally lack the skills to deal with problems of life in general. They were failures in interpersonal relationships and in occupational activities. The report states that "For many, arson may be the only thing they have tried in their life that yields relative success."[17]

The Bureau of Alcohol, Tobacco and Firearms, U.S. Department of the Treasury, in 1992 to 1993 sponsored a research project on the relationship between drug activity and arson. The U.S. Fire Administration, the International Association of Chiefs of Police, and agencies having fire investigation responsibilities in five cities also participated. The project was conducted by Virginia Commonwealth University in Richmond.

The study identified the following problem areas in arson detection, investigation, and reporting:

> (1) Unique characteristics of arson impede detection, investigation and reporting; (2) There is a lack of agency participation in reporting arson; (3) Lack of uniformity in measuring and defining arson variables impedes accurate statistical accounting of arsons; (4) Jurisdictional and organizational problems confound arson investigations; and (5) There is an immediate need for standardized and comprehensive arson investigation training.

The percentage of drug-related arsons in the five cities studied ranged from a high of 24 percent in Kansas City, Missouri, to a low of 13 percent in New Haven, Connecticut. Arsons represented 34 percent of all fires in both cities during the study period. The five-city average for drug-related arsons is 22 percent, thereby indicating that the drug-arson association is a definite factor in the arson-suppression picture.

The study recommends additional research on drug-related arsonists, which would in turn enhance prevention strategies.[18]

Another study illustrated the value of thorough fire investigations followed by vigorous prosecution. In this project, which was prepared for the Law Enforcement Assistance Administration, arson incidents, arrests, and convictions were analyzed in 108 cities over a four-year period. Cities ranking in the upper third according to arson arrest rates had 22 percent fewer arsons per 100,000 population than cities ranking in the bottom third; cities in the upper third, according to conviction rate, had 26 percent less arson.[19]

Arson Control Needs

The International Association of Arson Investigators and the National Fire Protection Association sponsored a symposium of Arson Control in the 1990s in Houston. Nationally recognized speakers from law enforcement, fire services, the insurance industry, and legislatures participated in this program. Several of their conclusions follow:

- Automatic sprinklers, public fire safety education, smoke detectors and code enforcement all play an important role in arson suppression.
- Training for fire investigators must be improved. In some cases the arsonist is more sophisticated than the fire investigators.
- Safety and health issues for fire investigators need additional study.
- Key members of state legislatures need to be made aware of arson problems and solutions.
- Public awareness of the arson problem is essential to effective abatement. Accurate data are of fundamental value in this effort.
- Wildland arson problems are becoming closely associated with suburban and urban areas as the trend toward construction of homes in woodlands continues.
- Some aspects of programs that address serial arsonists and those that relate to juvenile fire setters are similar. Interconnection responsibilities should be explored.
- The sheer volume of motor vehicle fires creates a gap in terms of thorough investigation. This gap needs to be addressed.[20]

Summary

The suppression of arson through the detection, investigation, and control of maliciously set fires has an important place in efforts directed toward the prevention of fire. Arson is increasing as a cause of fires in this country, and there is a growing awareness of the need to deter this kind of criminal action. Coordinated effort between police and fire agencies in the investigation of suspicious fires to bring about prosecution of the arsonist is needed in many communities. There is a need for more qualified fire investigators in the field and thorough training in investigative techniques to discover and preserve evidence of arson that will hold up in the courts. The crime of arson often goes unrecognized because of the lack of thorough investigation or because of failure to develop sufficient information to place an action in that category.

The model arson law, developed by the Fire Marshals Association of North America and adopted in the laws of many states, defines four degrees of arson with corresponding penalties. The term *willfully and maliciously* is a key factor in these provisions. To obtain a conviction of arson, a prosecuting attorney must be able to prove in court that the fire was in fact willfully and maliciously perpetrated. Some states have adopted the Model Penal Code of the American Law Institute, which differs in some respects from the model arson law.

It is helpful to the prosecution of a case if a motive can be established. Motives for arson range from insurance fraud, revenge, and cover-up of another crime to satisfaction of the abnormal urges of the vanity fire setter and the pathological fire setter.

A valid approach to the suppression of arson is a thorough investigation of all fires. No longer can the fire officer assume that a fire is accidental in origin just because the location or the owner is well-known in the community. Because motives for a deliberately set fire may not be readily apparent, no presumptions can safely be made.

A systematic examination of the fire scene is all-important in establishing the origin of the fire, whether accidental or not. The fire service personnel who arrive first should observe general conditions: people and vehicles in the vicinity, the color of smoke and flames, the presence of distinctive odors, the intensity and travel speed of the blaze—all of which can be clues for the investigation. If arson is suspected, a fire investigator is usually called to the scene. A trained investigator will look for any traces of incendiary origin of the fire, such as the use of flammable liquids or accelerants and evidence of tampering with equipment, such as gas valves left open on heating equipment or stove burners turned on. Photographs may also be taken at the scene. All evidence should be delineated in the fire investigator's report, and all physical evidence should be carefully preserved for laboratory examination and possible use in criminal court proceedings.

Arson is especially prevalent during periods of civil and economic unrest. Difficulties of investigation, collection, and preservation of evidence are compounded in fires that occur during these periods because of multiple fire calls, false alarms, and confusion and mob intervention at the scene.

Statistics of the FBI *Uniform Crime Report* show arson and malicious burning as a major crime problem in the country, and figures prepared by the National Fire Protection Association show that losses from incendiary and suspicious fires have increased alarmingly in the last two decades. This is a serious matter for the fire preventionist. The only way to combat it is to persist in efforts toward exposing and suppressing the crime of arson.

Notes

1. *Assembly Proceedings,* February-March 1638/39 (St. Mary's City, Md.), p. 72.
2. Betty Richardson and Dennis Henson, *Serving Together: 150 Years of Firefighting in Madison County, Illinois* (Collinsville, Ill.: Madison County Firemen's Association, 1984).

3. Steven R. Frady, *Red Shirts and Leather Helmets* (Reno, Nev.: University of Nevada Press, 1984), pp. 144, 145.

4. U.S. Department of Justice, Federal Bureau of Investigation, *Crime in the United States: Uniform Crime Report, 1992* (Washington, D.C.: U.S. Government Printing Office, 1986), pp. S1-S4 36-39.

5. *National Fire Protection Association Fire Report* (Quincy, Mass., 1976).

6. *Arson Facts, 1993-94,* (New York: Insurance Committee for Arson Control, 1994), p. 1.

7. *Model Arson Law* (Quincy, Mass.: National Fire Protection Association, 1931).

8. *Model Penal Code* (Philadelphia: American Law Institute, 1962), p. 152.

9. Arkansas Criminal Code, Little Rock, 1991.

10. Criminal Law, *Damage to Property: Arson* (Ottawa, Law Reform Commission of Canada, 1984).

11. *Los Angeles Civil Disturbance, April 29, 1992* (Los Angeles Fire Department, 1992), pp. 227, 229, 248, 250.

12. U.S. Department of Justice, p. 54.

13. "Fire Saver—Dealing with the Juvenile Firesetter in Memphis, TN" (Memphis Fire Department, unpublished paper) Memphis, TN pp. 1, 2.

14. *Michigan v. Tyler,* 436. U.S.499, U.S. Supreme Court, 1978.

15. *Michigan v. Clifford,* 52. U.S.4056, U.S. Supreme Court, 1984.

16. *State of South Dakota v. Jorgensen,* Opinion No. 13815, South Dakota Supreme Court, 1983.

17. Allen D. Sapp, Timothy G. Huff, Gordon P. Gary, David J. Icove, and Philip Horbert, *A Report of Essential Findings from A Study of Serial Arsonists* (Quantico, Va.: National Center for the Analysis of Violent Crime, 1994).

18. *Drug Related Fires in the United States* (Richmond: Virginia Commonwealth University, 1993).

19. Aerospace Corp. Survey and Assessment of Arson and Arson Investigation prepared for Law Enforcement Assistance Administration (El Segundo, Calif., 1976).

20. *Report on Arson Control in the 1990's: A Symposium* (Houston, TX: International Association of Arson Investigators and National Fire Protection Association, 1991).

CHAPTER 12

Fire Prevention Research

Until the 1970s there had been a limited amount of research in the field of structural fire prevention. On the other hand, there had been some excellent research efforts in forestry fire prevention.

The term *research* is defined as "careful, systematic, patient study and investigation in some field of knowledge, undertaken to discover or establish facts or principles."[1] A great deal of research effort has been expended in both structural and forest fire prevention during the last quarter of the 20th century.

The need for structural fire prevention research is exemplified by the experience of a Coast Guard petty officer who was assigned to waterfront fire prevention responsibilities in Seattle during World War II. During one week the assignment was to post fire prevention posters in all waterfront facilities in Seattle. The following week the assignment was to conduct fire prevention inspections of the same facilities. The question later arose as to which week was the most effective as far as utilization of time for fire prevention was concerned. Were they of equal value, or was one more productive than the other? Unfortunately, little research has been done to provide answers to such basic questions.

A revealing research project was conducted in the 1950s by the Ithaca, New York, fire department. The department was concerned with the great number of fires occurring in the city as a result of the use of portable kerosene heaters and decided to have feature articles written for the local newspaper to depict dangers inherent in the use of these devices. The articles were amply illustrated and were designed to discourage the use of this equipment except under safe conditions. They learned that the newspaper campaign was of practically no value in reducing fire losses.

As a result of this experience, a survey was conducted to determine the newspaper reading habits of the individuals with the portable heater problem. An analysis of the city's fire records showed that practically all of the fires involving portable heaters occurred in lower-income residences. Further research indicated that individuals residing in these residences had reading habits that were quite different from those of individuals with higher income levels. Most of the former group did not read the feature page of the newspaper but were concerned primarily with the help wanted page, the comics page, and in some cases the sports page. This research was responsible for a change in direction of the fire prevention program to get a message across to the people who were most directly part of the problem. This is an example of local research that can be most helpful to an fire department.

U.S. Forest Service Fire Prevention Research _____

The Forest Service of the U.S. Department of Agriculture has carried out a number of excellent fire prevention research projects. Some of these were conducted by Forest Service personnel; others were conducted by university personnel under contract to the Forest Service.

Smoking Regulations Signs

The Pacific Southwest Forest and Range Experiment Station of the Forest Service conducted a research project pertaining to no-smoking signs.[2] In many of the Southern California mountains, there is a prohibition against smoking while traveling in a vehicle through the area. These restrictions apply to some high-speed multilane highways. Signs are provided at the entrance to these restricted zones; in some cases areas are provided in which a motorist may stop and smoke. A sign that showed a smoldering cigarette with the word NO was compared to a sign with a picture of a smoldering cigarette with no wording but with a slash across the cigarette. The signs were very effective in attracting the eye of motorists. Eighty-five percent of those passing the one with the word and the picture recalled the message. Seventy-three percent of those who saw the sign with no wording recalled the message. This indicates the desirability of illustrated signs with a very limited amount of wording as a public warning.

A similar survey was done with the signs for indicating areas where motorists might stop and smoke. Again it was found that a sign with OK on it was superior to a sign that merely showed a lighted cigarette with no wording whatsoever and without the slash mark. An interesting feature of the latter survey was the tendency of people stopping at the smoking area to steal the signs. A great number of the signs were apparently removed by youthful sign collectors.

The Pacific Southwest Station in Berkeley also conducted a survey regarding knowledge of smoking regulations in the national forests of Southern California.[3] They were interested in determining the accuracy of people's understanding of forestry regulations. They found that individuals living close to the forests were generally more familiar with the regulations than were those who lived at distant locations. They also found a direct relationship between frequency of visits to the forest and possession of correct information. Those who visited the forest ten or more times understood the regulations almost twice as well as those who were traveling in the forest for the first time. The researchers found that approximately one of four smokers traveling in the forest does not understand prevailing smoking regulations. This clearly indicates the need for disseminating specific information.

Children and Matches

The U.S. Forest Service, in cooperation with the University of Southern California and the California Division of Forestry —now the Department of Forestry and Fire Protection—also conducted research into "children with matches" fires.[4] This research was conducted to determine problems with and possible solutions for these fires. Approximately three-fourths of the fires attributable to juveniles were started by children 10 years of age or younger. Twelve percent of the 484 fires studied were started by children under five years of age, while 32 percent were started by children between five and seven. Children 8 to 10 caused 28 percent, while the remaining fires were caused by children 11 or older. Ninety-two percent of the offenders were boys.

The direct result of this research was a realization that fire prevention education should include preschool and early elementary children if there is to be abatement of the fire setting problem. Very young children might be most effectively reached through popular television programs for their age level. The research bears out the need for interchange of information among investigative agencies handling the fire problem within a given area.

Public Awareness of Fire Prevention Importance

An earlier study by the Pacific Southwestern Forest and Range Experiment Station addressed itself to the fire prevention knowledge and attitude of the residents of Butte County, California.[5] This county was selected because of the high percentage of the population that goes into the wild lands of the county at some time during the year. All social and economic factors regarding the surveyed residents of Butte County were analyzed. These included age, marital status, education, job status, type of work, sex, race, present and previous places of residency, or migratory status. The survey indicated that fire prevention knowledge was higher among individuals who used the wild land either for pleasure or work on a frequent basis. It also indicated a definite relationship between socioeconomic variables and experience with forest fires. It noted that frequent users scored best in both knowledge and attitude scales. There was a strong and positive feeling toward fire prevention on the part of most respondents. This feeling was intensified by personal experience with wildfires and the extent of experience in wild land areas.

There were also noticeable groupings of individuals who are frequent users of the wild lands who had little knowledge of or interest in fire prevention problems. These included high school dropouts. Contrary to general opinion, females were equally represented in this group. A number of other things were considered in this survey, including an analysis of fire prevention information received through television and newspapers.

Research in Fire Prevention Promotion and Enforcement

The Forest Service also conducted a survey in Southern California to determine the best means of promoting fire prevention.[6] They found that a successful fire prevention campaign contains two elements: It should teach *why* individuals must be careful with fire, and it should also teach *how* we should be careful with fire. An attempt was made to determine what the public already knows about fire prevention. Researchers went into the forests to study fire prevention measures. They interviewed individuals who had been in the forests and questioned them about their fire prevention practices.

They found that men generally scored better than women in fire prevention knowledge. They also found that people in the 25 to 50 age group had the most correct information on fire prevention and that those in the under 25 age group generally lacked information. In the over 50 age group, the problem was with misinformation.

The study further revealed that individual contact taught the most, followed by information contained in written material. The third most important means of conveying fire prevention information was through radio and television, which were useful in conditioning attitudes. Individual contact was found to be useful in pointing out specific fire prevention procedures. Written material was also found to be helpful in the latter category.

Enforcement of Fire Laws

The same study by the Forest Service also addressed the problems of fire law enforcement. The researchers interviewed individuals who had been charged with violations of fire prevention regulations. They also interviewed individuals who had followed favorable practices in the use of campfires in forested lands. This study was quite comprehensive and included a considerable amount of review of case studies as well as interviews with appropriate individuals. The results of this survey were revealing and gave indications of problem areas that relate to structural as well as forest fire prevention.

The survey revealed that the average violator in the forested area was an honest, intelligent individual who was aware of the importance of fire prevention but who had not read or understood the specific regulations. The survey further indicated that the majority of people believed that fire prevention regulations are not generally known and recognized and that fire prevention laws and regulations are as important as traffic regulations. Most of those interviewed stated that they would observe both types of regulations equally. Most of the nonviolators interviewed said that they saw the need for more enforcement of forest fire prevention regulations; less than one-half of the violators interviewed believed that less enforcement should be practiced.

A survey of the practices of the violators revealed that most did not realize they were doing anything incorrectly. They were following practices they thought were safe but that turned out to be violations of fire prevention regulations.

Both violators and nonviolators were asked the question, "Have you received any traffic citations in the last two years?" In answering this question in the affirmative, violators outnumbered nonviolators three to one. This indicates the possibility of a close relationship between individuals who violated fire prevention regulations and those who violated traffic regulations and laws. It may further indicate that certain members of the public are indifferent to regulations that affect them directly and thus choose to ignore these requirements.

Forest Service research projects have indicated that there are three component areas of fire prevention effort: education, enforcement, and engineering. At times it is difficult to draw a line between these segments.

Another phase of the study examined enforcement efforts by individuals assigned such responsibilities. The assignment of personnel to fire prevention enforcement duties does not necessarily mean that prompt enforcement will be effected. The survey noted the need for full review of enforcement policies relating to fire prevention requirements and a review of the uses of the laws and regulations in effect. Certainly, there is little value to laws that are on the books but that are not enforced. This portion of the survey recognized that the existence of a regulation does not necessarily mean that full compliance will be achieved.

Media Uses in Fire Prevention Promotion

The study included a detailed evaluation of the "senders," who might be defined as the sources of information in the fire prevention field, as opposed to the "receivers." This included a detailed survey of all media used for disseminating fire prevention information.

The study surveyed the degree of credibility placed on the senders by the receivers. Some individuals are not inclined to believe information on television or in a newspaper. Such prejudices were evaluated so that people responsible for publicizing fire prevention might be guided in using the proper media. The program also included a survey of media employed for fire prevention efforts with relative values assigned. An evaluation was made of the credibility of media, generally without specific reference to credibility of the particular receivers who were interviewed. This involved a general survey of public information to determine appropriate use of the different methods. The methods were studied from the standpoint of applicability to the public. This portion of the survey was primarily related to forest users who were considered as "the public" for the purpose of the survey.

The messages given out in fire prevention education efforts were also studied in detail for composition and general attractiveness. Message content was analyzed, with emphasis on the semantics of the message. The words *water shed,* for example, were found to connote to many members of the public a structure in which water is stored. This indicated the need for the use of a word having more clear meaning to the majority of people.

The survey checked the use of the words *closed area.* It found that this term also had little meaning to the average person who was not fully familiar with forestry procedures. A number of people were not sure of the meaning of the term *limited access* and could be easily misled by a sign bearing it. The public had varying opinions of the meaning of *highway* and *parkway.*

Los Angeles "School Burning" Tests _____

An example of a research project that was directly related to a particular fire was the extensive "school burning" project carried out in Los Angeles.[7] This project was carried out to determine the best procedure to obtain fire safety in schools subsequent to the Our Lady of Angels School fire, which occurred in Chicago in 1959. Financing for the tests was provided to a great extent by the Ford Foundation; however, the Los Angeles Fire Department, the National Fire Protection Association, and a number of other organizations gave freely of time and effort.

Applied Physics Laboratory Research on Fire Deaths _____

Another example of research in the fire prevention field is the fire fatalities project carried out starting in 1971 by the Applied Physics Laboratory of Johns Hopkins University.[8] This federally subsidized project involved a detailed analysis of fire deaths in Maryland by research personnel. These researchers reviewed available information on fire deaths and attempted to determine direct causes as well as other pertinent information surrounding these deaths. In each case all possible sources of information were contacted, and the individuals from the laboratory developed trends from the available data. A study was made of the deaths occurring in various age groupings, and an effort was made to relate fire causes to this information.

The survey was carried out in cooperation with the Maryland State Fire Marshal's office, the Baltimore City Fire Department, and the Maryland State Medical Examiner's office. In addition, county fire marshals' offices in Prince George's, Anne Arundel, Montgomery, and Baltimore counties participated in this project.

Researchers responded to the fire scene and made a detailed analysis of the premises. Structural considerations were noted, and an effort was made to analyze the reasons for fire spread and for any unusual occurrence in connection with the fire. This physical survey of the fire scene was generally made in connection with a repeat visit to the scene by a representative of the investigating agency having jurisdiction. The researcher proceeded to review the known information regarding the fire and to interview witnesses.

The program included detailed autopsies on all fire fatalities. An autopsy was conducted to determine any unusual features, such as alcohol in the blood or evidence of narcotics use. A great effort was made to determine blood levels of carbon monoxide and cyanide.

The first part of the survey addressed the problem of "children and fire fatalities." The report indicated that fire-related literature has long created the impression that children four years of age and younger have a high fatality rate in fires. For this reason a survey was undertaken to understand the nature of the fires causing these fatalities. Questions arise as to the causes of the fires, reasons for lack of evacuation and rescue, and other related issues. The study considered children of 17 years of age and younger. A total of 321 fatalities were included in the survey. Of these fatalities, 86 percent (278 victims) were in residential occupancies including homes, apartments, and mobile homes. These 86 percent of the fatalities occurred in 225 fires. In 35 fires there was more than one fatality, with a total of 85 victims involved.

In 69 of the residence fires surveyed, at least one person was known to have escaped. This point is often not included in surveys of fire fatalities, and it does raise some interesting facts for consideration. This indicates that at least one person escaped in approximately 30 percent of the residential fires surveyed. However, this cannot be considered a fully valid figure, because escapes are not routinely recorded in reports of residence fires.

In the six-and-under age category, there were 47 fires involving 68 fatalities. Of these fires, 17 had multiple fatalities involving 38 deaths among children in the one-through-six age group. Approximately 50 percent of these fires had escapees. The survey further noted that the predominant time for fatal fires in the one-through-six age group is between 6 A.M. and 12 noon. In all but 1 of the 14 fires noted, the deaths occurred as a result of fires between 8 A.M. and 12 noon. In one year studied, all 5 of the fatal fires involving children playing with matches occurred between 8 A.M. and 12 noon. Only 2 children died as a result of fires between midnight and 6 A.M. This seems to be contrary to popular conceptions about the times of fires.

The survey also indicated that the bedroom is the most dangerous area of the house for fires in which children are killed. There seems to be strong evidence in the survey that children play with matches in their bedrooms or in adjacent rooms. In the latter case, the child may run back to the bedroom after having started a fire in some other location.

In fire deaths among children of six and under, the main cause of the fire is playing with matches. This indicates a strong need for positive steps in an effort to reduce this fire cause. If the survey were replicated today, lighters would probably as much a problem as matches.

There is also an indication that the lack of central heating is a primary fire cause among the age group of up to six year of age. This includes the use of kerosene space heaters and other portable heating appliances in residential occupancies. Some jurisdictions have banned the sale of devices of this kind. Both of the foregoing causes indicate the value of fire prevention education as a means of reducing fire problems among these age groups.

The survey on the second age group, ages 6 to 17, was chosen to include school-age children. The surveyor thought that fire safety programs might be addressed to this age group and that a separate category would give an opportunity to evaluate the value of fire prevention programs. The fact that teenage children serve as baby-sitters was also considered in the development of this category. The study included 30 fires with 38 fatalities in that age group. This included 10 fires with multiple deaths involving 18 children. In 14 fires involving children of this age group, there were escapees.

The survey also indicated a similarity between the causes of fires within both age groups. This is especially noticeable in the consideration of fires resulting from the lack of central heating systems.

The survey also addressed the problem of unattended children. In the sample fires studied, there were 22 fatalities in 11 fires in which unattended children were involved. The age group of six and under had the largest number of fatalities. The time period of 6 P.M. to midnight is the most dangerous as far as unattended children are concerned. This may indicate that the time of departure of parents and guardians is geared to their social activities, which usually take place between 6 P.M. to midnight. This is also a time when many people are employed on second jobs.

The survey indicated a great need for public awareness of fire hazards. This need must be met by an effective educational program in fire safety, possibly as a part of an overall public safety program in the public and private schools. There is also a strong indication that teenagers should be included in such a program.

The survey indicated that matches are the primary ignition source in fires involving children. All children who died in fires that were attributable to matches were six years of age or younger. The fires in all cases occurred between 8 A.M. and noon. Only one of the fatal fires in the study was attributable to smoking.[9]

The second part of the survey, which addressed adult fire fatalities, indicated that approximately 80 percent of the fatalities involved carbon monoxide inhalation. Approximately 59 percent of the bodies examined had sufficient carbon monoxide for it to be classified as the cause of death. Alcohol was a factor in over 40 percent of the cases. Advanced age was also a contributor to death. Preexisting heart conditions contributed to 12 percent of the deaths.

In only a small number of deaths were burns considered to be sufficient to have caused death. This finding is somewhat contrary to the usual concept of fire deaths. The burn deaths were generally a result of fires caused by smoking or other careless use of smoking material. This was true in practically all of the fatal fires included in the survey. The researchers have noted that there was a relatively low percent of narcotics use apparent on the part of individuals who died as a result of the fires surveyed.

This survey found that the primary time for fatal fire is between midnight and 7 A.M. It also noted in correlation of data from the medical examiner's office and from the study of the fire scene that fires as a result of smoking usually occur during those times. This pertains to fires in which careless smoking has been found to be the primary cause of death.[10]

The survey encompassed 315 fatal fires in which 421 people were killed. Fifty percent of these fires occurred as a result of careless smoking. Other causes included misuse of flammable liquids, arson, heating equipment, electrical, matches, candles, and a large miscellaneous group. Soot was found in the lower respiratory tract of 90 percent of the cases analyzed.

The percentage of fire fatalities caused by smoking was the highest in July and May and the lowest in June and December. The most likely hour for a fatal fire was 4:00 A.M. followed by 3:00 A.M. and 5:00 A.M. The least likely hours were 10:00 and 11:00 A.M.

The bedroom was the most likely area of origin, with 49 percent of the fires having originated in that room. The living room was the room of origin in 22 percent of the fires, while 7 percent originated in the kitchen and 6 percent started in the basement.[11]

A companion study was carried out by Applied Physics Laboratory personnel to assess the potential impact fire safety systems could have on reducing casualties and property damage in dwelling places. The area of the survey was expanded to include the District of Columbia.

Three separate systems were considered: a smoke detection system; a smoke detector system with an automatic alarm to a fire alarm center; and an automatic fire suppression or sprinkler system designed for residential use.

Smoke detectors installed in conformance with National Fire Protection Association Standard No 74 (now 72), were considered to be smoke detection systems in the survey.

At least 101 of the 114 fatalities could have been avoided and at least 110 of the 119 injuries could have been avoided had a smoke detection system been installed, the surveyors found. A suppression system would have made an even greater contribution with a potential saving of 105 lives and a greater amount of property. It was estimated that 103 lives would have been saved with the automatic alarm connection coupled with smoke detectors.[12]

Another fire study of major consequence was carried out in connection with toxic gas accumulations in building fires.[13] The buildup of carbon monoxide and carbon dioxide, as well as the depletion of oxygen, in upstairs fires was studied. The study found that in a fire situation in a two-story dwelling, dangerous concentrations of carbon monoxide were present in a bedroom on the second floor of the building in approximately 10 minutes. Fuel burned for rapid fire buildup in this test consisted of a wooden crib. In the situation where a slow buildup fire involved an overstuffed chair, dangerous conditions did not become noticeable until approximately 20 minutes after the start of the fire. This information was useful from a fire prevention and planning standpoint and could be useful in the development of equipment for use in detecting fires. Toxic accumulation tests were conducted by Applied Physics Laboratory personnel in cooperation with personnel of the University of Maryland fire protection engineering department.

The Alabama State Fire Marshal's Office prepared a detailed report on fire deaths in that state for one year. Each incident was analyzed, and interviews were conducted to ascertain conditions surrounding the fires. Post mortem examinations were also performed as a part of this research project.[14]

The Alabama project revealed many similarities with research conducted by Johns Hopkins University. An additional factor that was considered in the Alabama review was that of marital status. A high percentage of fire victims were found to be single, widowed, or divorced. The project also analyzed types of public fire protection service available at the fire site. Well over half of the fatalities occurred in localities protected by career fire departments. The larger cities were well represented in the total fatality summary. An update of this report 12 years later indicated similar findings including a high percentage of alcohol use by victims, but very little use of narcotics.[15]

High-Risk Area Research _____

New Orleans was the site of two meaningful research projects in fire prevention carried out with federal support. The first study was conducted by Planning and Research Associates, Inc., of Baton Rouge, with the assistance of the New Orleans Fire Department. It was entitled *The Human Factor in High Fire Risk Urban Neighborhoods.*[16] The second study was entitled *Fire Prevention Program Planning in a Social Action Perspective: An Exploratory Study of Knowledge, Attitudes, and Leaders in a High Fire Risk Area of New Orleans, Louisiana.*[17] This survey was also conducted by Planning and Research Associates, Inc., with the assistance of the New Orleans Fire Department.

Both studies took place in a high fire risk urban residential area in the central part of New Orleans. The area had around 25,000 inhabitants, of whom 91 percent are black. The predominant housing unit is the duplex, although there are some single-family and high- and low-rise housing projects in the area. Ten percent of the housing units were occupied by their owners at the time of the study.

The goal of the first study was to determine the economic, cultural, and demographic variables related to fire occurrence in a high fire risk urban residential area. The study resulted in the following summary of findings:

1. Preliminary investigation determined that the types and kinds of fire in the study area were similar to those found in high risk areas in other cities.
2. Interviews with firemen and residents indicated that fires were primarily caused by careless behavior, such as smoking in bed, placing

flammables near open flames, and leaving food unattended on stoves; and by tolerant attitudes toward trash and debris and behavior of juveniles and vagrants.

3. Analysis of the data obtained suggested that the human factor, expressed in terms of a subculture, would have to be considered in planning programs of fire prevention. Fire proneness is the result of these specific subcultural characteristics:
   a. lack of a strong spirit of community,
   b. lack of fire safety training,
   c. lack of practice of fire precautions,
   d. lack of fire fighting facilities, and
   e. an attitude of fatalism toward fire.[18]

The study found that many children in the area use fire as a plaything and that this practice is not strongly discouraged by parents and other adults. It also noted that the area under study tended to attract all types of vagrants, many of whom frequented vacant buildings. These individuals seemed to be more readily accepted in the study neighborhood than they would be in many other areas.

Preventive programs for high fire risk areas will need to be especially designed, the survey indicates. There is a need to change ways of thinking about fires in such areas; there is also a need to identify and utilize proper communication channels to reach the target population, according to the survey report.

The second study attempted to determine what social action could be initiated at the local level to reduce the rate of fires and fire losses in residential areas with high-risk characteristics.[19]

Research personnel found that one-third of the residents interviewed in the target area had experienced a fire in their dwelling at one time or another. Of these, half had experienced a fire within the previous 10 years. Residents experiencing fires seem to have "learned a lesson" from their particular fire but not a general lesson in fire prevention. They did not gain a full appreciation for the broad concepts of fire prevention but only gained an appreciation of preventive measures relating to their personal fire experience, whether it be careless use of smoking material or leaving food on the stove unattended.

Personnel assigned to the survey found that there is often a distrust of outsiders in neighborhoods of this type. There is also, they found, a fear of involvement and distrust of "programs" as well as a casual attitude toward fire prevention. They concluded that the conditions found in New Orleans were generally similar to those of high fire risk areas in other cities.

Specific conclusions of the second survey include a finding that the residents of the area studied have a strong awareness of fire as a problem. They also found a strong belief among the residents that the fire department should take the responsibility for fire prevention programs at the city level. They believed that neighborhood programs in fire prevention could be performed by other public agencies as well, such as schools and social service agencies.

The survey found that the mass media, especially television and radio, have a potential for reaching residents with a fire prevention message. The leadership structure in the study area is complex, and this force is not as apparent as in more cohesive neighborhoods. Local leaders, it was found, are primarily church leaders, government employees, and elected officials.

Surveyors contacted the area leaders and found that these individuals were well aware of the fire problem and the need for fire prevention programs. These individuals indicated an apparent willingness to participate in such activities.

The New Orleans projects gave fire prevention personnel some excellent information on which to base programs for high fire risk areas. The results can be of value in designing programs for such areas.

National Institute of Standards and Technology Research _____

Research activities of the Building and Fire Research Laboratory of the National Institute of Standards and Technology include multidisciplinary projects directed at the development of predictive capabilities and other tools needed by the fire safety community to reduce fire losses and costs. The Laboratory's technical program is carried out by five divisions. The divisions that relate to fire safety are described below:

> The Fire Safety Engineering Division develops methods to predict the behavior of fire and smoke and to assess ways of mitigating their impact on people, property, and the environment.
> The Fire Science Division develops metrology for fire research and understanding of the scientific and engineering aspects of fire phenomena.[20]

Examples of fire safety–related research activities conducted by these divisions include the development of an occupant fire evacuation plan for board and care facilities. This project entails a study of the application of the Fire Safety Evaluation System to these facilities, a review of factors surrounding significant fires in board and care occupancies, and an evaluation of egress conditions with an engineering analysis using computer models that simulate fire environment, models predicting people movement, and models predicting human debilitation from heat and chemical loading.

A risk analysis for the fire safety of airline passengers and continuing work on development of performance criteria for improvement of the fire resistance of upholstered furniture are among the many activities of the Laboratory. New projects are undertaken based on governmental direction or industry demand.[21]

Several projects of the Laboratory have already had major impact on fire safety. One is the development of a standard for flammability of children's sleepwear. Before the development and promulgation of this standard, now administered by the Consumer Product Safety Commission, as many as 300 children were killed and thousands injured each year. The standard became effective in the early 1970s.

Another major impact project is the development of the standard for cigarette ignition resistance of mattresses. Cigarette ignition of mattresses has caused thousands of fatalities through the years. This standard, also administered by the Consumer Product Safety Commission, is reducing the incidence of cigarette ignition of mattresses.[22]

Personnel of the Laboratory are active in many standards-making organizations. This is one of the methods for transferring technology developed by the Laboratory into useful applications. A summary of the goals of the Laboratory is contained in Chapter 10.

Canadian Research Activities _____

The Institute for Research in Construction of the National Research Council of Canada was established to serve the needs of Canada's construction industry. This industry is Canada's largest, representing almost 20 percent of the gross national product.[23]

The National Fire Laboratory of the Institute is responsible for conducting research aimed at reducing life and property losses by fire. The Laboratory was established as a direct result of a major ship fire in the port of Toronto. Priority is given to subjects related to the National Building and Fire Codes of Canada and to assistance to the building industry through research into fire reaction of building materials and assemblies.

The Canadian fire research facility located near Ottawa is probably the finest such facility in the world. It includes a 10-story tower in which high-rise fire safety studies can be conducted. It has been used to conduct evacuation research for such buildings.

Research activities of the Laboratory include fire risk assessment, costs of fire safety including public protection, effectiveness of structural fire protection including fire barriers, and high efficiency water-based fire suppression systems. Other aspects of fire protection are also studied.

Figure 12-1   Canada's National Fire Laboratory is equipped to handle a wide range of fire safety testing. (National Fire Laboratory, Ottawa.)

Code Enforcement Research

The Urban Institute, the National Fire Protection Association, the National Science Foundation, and the U.S. Fire Administration combined forces to conduct extensive research into the area of fire code inspection practices.[24] Experts have long thought that fire code inspections result in fewer fires, lower fire losses, and fewer civilian casualties in properties covered by fire codes. They have also thought that fire codes prevent hazards and deficiencies that cause fires.

The study identified the fires and civilian fire casualties that may be prevented by inspection and measured the number of such fires and casualties. It also identified potentially important characteristics of fire code inspection practices and measured differences in practice from city to city. A representative group of 11 large cities was selected for in-depth study. Data from 6 other cities and 1 county were used for comparative purposes. Incendiary and suspicious fires were excluded because there is limited potential for preventing them through inspections.

Cities selected for the survey were Charlotte, North Carolina; Columbus, Ohio; Denver, Colorado; Houston, Texas; Indianapolis, Indiana; Jacksonville, Florida; New Orleans, Louisiana; San Jose,

California; Seattle, Washington; Tucson, Arizona; and Wichita, Kansas. The fire code inspection practices analyzed were divided into four groups: measures of inspection frequencies; inspector preparation (training and experience); motivational practices, including techniques used to ensure compliance; and practices, including plan reviews, aimed at controlling building construction features related to fire safety.

The comprehensive report of this project contains several conclusions. Because these conclusions contain some excellent suggestions for fire safety inspection program management, they are included in summary form:

1. Cities that annually inspect all or nearly all inspectable properties appear to have substantially lower fire rates than do other cities. Cities in which a substantial share of the inspectable properties go several years between inspections, or are not regularly inspected at all, tend to have higher fire rates.

2. Cities that use fire suppression companies for a large share of their regular fire code inspections appear to have substantially lower fire rates than do cities that use full-time fire prevention bureau inspectors exclusively. The probable reason is that cities using full-time inspectors exclusively often do not have sufficient personnel to make annual inspections of all inspectable properties, whereas cities that use fire suppression companies usually have the needed personnel.

3. Cities that assign inspection responsibility by geographic districts appear to be more successful in covering all inspectable properties. In the district-based approach, city blocks or streets are covered each year, and therefore each inspection cycle provides a new opportunity to locate an inspectable property that was not spotted before, either because it was not there in the previous cycle or because it simply was not noticed before.

4. There was no evidence that differences in fire rates are sensitive to other differences in the inspection practices examined: i.e., techniques used to prepare inspectors, to ensure thorough inspections, to persuade owners and managers to comply, to enforce compliance when voluntary compliance is not possible, or to control building features. The word *technique* is important here because action is being assumed in all cases; the variable is merely in methods or "techniques" used to achieve the goal.

5. Despite the almost universal belief that selling fire prevention is the key to success in obtaining compliance, inspectors were given little or no training or policy guidance on persuasive techniques.

6. Although the evidence did not show that any particular persuasive techniques were associated with lower fire rates, there was

some evidence that the motivational and persuasive impact of inspections is more important than the direct removal of hazards. In inspectable properties in all communities examined, fires resulting from careless or foolish actions or electrical or mechanical failures (which are sometimes because of poor maintenance) greatly outnumbered fires resulting from visible hazards that inspectors are likely to remove directly. Fires of the first kind (which tend to be preventable primarily through educational and motivational efforts) typically constituted 40 to 60 percent of all fires in a particular community, while fires of the second kind typically constituted 4 to 8 percent of all fires. (The remaining fires were of incendiary, suspicious, natural, or undetermined cause.) The most important difference in motivational and educational effects thus appears to be whether inspectors get around each year; if they do, they will have more effect than inspectors who are rarely if ever seen.

7. Deficiencies in building features and the absence of automatic fire protection equipment were factors in most incidents involving 10 or more civilian deaths in inspectable buildings in the United States. Fire departments might well consider giving priority attention to building features most often implicated in fire involving large numbers of casualties. The lack of large loss of life in school fires since the early 1960s may be the result of the nationwide crackdown on school fire safety following the Our Lady of Angels fire in Chicago.

Study of Canadian Fire Safety Systems _____

References have been made in other chapters to the report *America Burning,* published by the National Commission on Fire Prevention and Control in the United States. This 1973 report analyzed fire problems and recommended a number of solutions.[25] The report was developed by a full-time staff with the guidance of the Commission by the president. The National Research Council in Canada, as a result of a petition from the Association of Canadian Fire Marshals and Fire Commissioners and the Canadian Association of Fire Chiefs, established a committee that recommended a comprehensive study of fire safety in Canada. Staff personnel traveled the country and interviewed people responsible for fire safety in both public and private sectors. They gathered a great deal of statistical data regarding such diverse subjects as "leading factors and lives lost through fire"; "percent of budget spent on prevention"; and "comparative fire loss in sprinklered buildings." Both public and private sector responsibilities and fire safety were analyzed in detail. Questionnaires were sent to municipal administrators and others responsible for fire safety. Results were analyzed, and recommendations were subsequently developed.

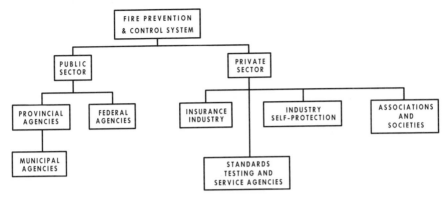

Figure 12-2 Chart in "Study on Fire Prevention and Control Systems in Canada," which shows that country's fire control system. It is very similar to systems in place in the United States.

The report, entitled *Study on Fire Prevention and Control Systems in Canada,* contains a number of specific recommendations, some of which imply criticisms of systems existing at the time of the report.[26]

Recommendations relating to fire prevention and education programs are noted below. They have application in both the United States and Canada:

1. The provinces and territories should establish a vigorous and competent program in fire prevention enforcement and public education as a combined effort of the province and its municipalities with particular emphasis on life safety. This should include:
   a. Fire safety education by means of established courses in the primary, secondary and post-secondary levels of the public education system.
   b. Education of the public on a regular basis through the media and other available means.
   c. Qualified advisory services to builders, industry and business.
   d. Examination of and advice on new building plans, inspection for code enforcement, advisory and educational visits to all buildings annually or more often as statistics indicate a need, utilizing Fire Marshals' or Commissioners' staff and municipal fire department personnel.
   e. Encouragement of municipal fire departments to greatly strengthen their role in all the foregoing activities, in addition to provision of technical assistance, training and any legislation changes necessary. To facilitate this there should be a major shift of firefighting into the dual suppression-prevention role.
   f. Provision of legislation for administration and enforcement of testing and maintenance programs by qualified personnel for fire detection, alarm, control and extinguishing systems in buildings.

An appendix to the report brings out the contrast between fire control and law enforcement as related to interjurisdictional cooperation. In Canada, as in the United States, there is a federal law enforcement focus. In fact, the federal force, the Royal Canadian Mounted Police, carry out local police functions in many parts of Canada. Training and research activities are also carried out at the national level.

Fire safety issues may be considered in research activities addressing segments of socioeconomic considerations. One such project was related to Indian conditions in Canada. This project, which was conducted by the Minister of Indian and Northern Affairs, Canada, addressed social conditions of Indians on and off reserve lands (known as reservations in the United States).[27] About 65 percent of Canada's Indian population lives in rural or remote communities, compared to 25 percent of the national population; 70 percent live on reserves.

The rate of violent deaths, including fires, for Indians is more than three times that of the national population. This situation may be attributed to lower quality of housing, substandard heating systems, crowded conditions, and scarcity of fire protection services. Only 30 percent of the reserves have adequate fire departments.

Canadian officials have been able to use information from this research project to develop fire safety procedures and programs for Indian reserves. The "Learn Not to Burn" program of the National Fire Protection Association has been tailored to include Indian characters as part of the government's effort to improve fire safety in this high-risk group.

National Fire Academy Research

Students attending the Executive Fire Officer Program at the National Fire Academy in Emmitsburg, Maryland, are required to prepare and submit an applied research project. Abstracts of the following two projects represent applied research that contributes to the advancement of fire safety:

A Study of Structure Fires and Fire Sprinkler Systems in U.S. Fire Stations

Fire stations are not immune from the threat of fire. Historical and descriptive research methodologies were used to answer these questions:

1. How many structure fires occur in U.S. fire stations?
2. What is the most common area of origin for U.S. fire station fires?

3. How many fire stations have fire sprinkler protection?
4. Would a properly installed fire sprinkler head reduce the fire damage losses in U.S. fire station fires?

Published literature and unpublished data were reviewed. A survey of 235 U.S. career fire departments serving less than 100,000 people was conducted.

The survey found 4.92 structure fires per 1,000 fire stations per year, compared with a national estimate of 3.46 structure fires per 1,000 fire stations per year. The kitchen/cooking area was found to be the most common area of origin for fire station fires. Partial and/or complete fire sprinkler coverage was found in 16.5% of the 915 fire stations surveyed. Fire damage losses would have been reduced in one-half of the fire station fires if there had been one properly installed fire sprinkler head.

A proposal to the *Standard on Fire Department Occupational Safety and Health Program—NFPA 1500,* requiring partial fire sprinkler protection in fire station kitchens of career fire departments was drafted. The Boca Raton Fire Department sought funding for implementing this level of protection in their existing fire stations. The study further recommended that all new fire stations should have complete fire sprinkler protection.[28]

Wood Shake Shingles ... the 800-Year-Old Problem

The purpose of this paper was to develop data supporting a ban on wood shake roofs in the San Miguel (Calif.) Consolidated Fire Protection District. Part of this data included a brief history of wood shakes, their contribution to conflagrations, and the persistent denial of same by the wood shake industry. This history began with untreated wood shakes, touched briefly on spray-treated shakes, and concluded with the latest, pressure-treated (also called pressure-impregnated) wood shakes.

The problem was our lack of an ordinance which would ban wood shake roofs. There existed many homes built within, or adjacent to, our wildland/urban interface that already had wood shake roofs, and the potential existed for many more. In fact, nearly all our available area of growth fell into the interface areas.

The findings indicate that no wood shake, treated or not, can be relied upon to provide adequate fire retardant capabilities in major wildfires or other conflagration-type situations. There are certainly claims to the contrary by the wood shake industry, but their test methods and subsequent findings were found to be questionable. History disproved old "safety" claims by the industry; current claims have already been challenged.

Because of this history, and our own climate, topography, seasonal Santa Ana (hot, dry, accelerated) winds and other conditions, it was recommended to push for a complete ban on wood shake roofs, pressure-treated roofs included.[29]

Summary _____

Although a considerable amount of research has been done in the general field of fire prevention, only a limited amount has been directed toward structural fire prevention.

Research programs carried out by the U.S. Forest Service have been directed toward efforts to prevent forest fires. Experiments have been made with different kinds of roadside fire prevention signs to determine their relative effectiveness. A similar project was designed to test the value of different kinds of no-smoking signs and to determine public understanding of smoking regulations. Other research studies have been made of fires caused by "children with matches," public awareness of fire prevention importance, public image of Smokey Bear and its effect on fire prevention, fire hazard inspection procedures, methods of fire prevention promotion, and problems of fire law enforcement.

The Applied Physics Laboratory of Johns Hopkins University has engaged in extensive research in many areas of interest in fire prevention. Among these efforts are surveys of fire deaths and analysis of fire death statistics.

Among areas in which additional research is needed is the problem of fires caused by children. More information is needed about the sociological and psychological factors that cause some children to set fires, and research may well point up the need to redirect fire prevention education programs for young children.

Researchers are leading the way to some scientific answers to the question of how to prevent fires, and recent trends indicate that research will become increasingly important in the future directions of fire prevention.

Notes _____

1. *Webster's New World Dictionary,* Second College Edition (New York: Simon & Schuster, 1982).
2. William S. Folkman, "Signing for the 'No Smoking' Ordinance in Southern California," U.S. Forest Service Research Note, PSW-71 (Berkeley, Calif.: Pacific Southwest Forest and Range Experiment Station, 1965).
3. William S. Folkman, "Motorists' Knowledge of the 'No Smoking' Ordinance in Southern California," U.S. Forest Service Research Note, PSW-72 (Berkeley, Calif.: Pacific Southwest Forest and Range Experiment Station, 1966).
4. William S. Folkman, "'Children-with-Matches' Fires in the Angeles National Forest Area," U.S. Forest Service Research Note, PSW-109 (Berkeley, Calif.: Pacific Southwest Forest and Range Experiment Station, 1966).

5. William S. Folkman, "Residents of Butte County, California: Their Knowledge and Attitudes Regarding Forest Fire Prevention," U.S. Forest Service Research Paper, PSW-25 (Berkeley, Calif.: Pacific Southwest Forest and Range Experiment Station, 1965).

6. Lloyd M. LaMois, *Research in Fire Prevention,* Second General Session, Third Annual Governor's Statewide Conference on Fire Prevention (Annapolis, Md., 1961), pp. 37–43.

7. *Operation School Burning* (Boston: National Fire Protection Association, 1962).

8. B. M. Halpin, "Sample Survey of Maryland Fire Fatality Data," National Science Foundation RANN Program/GI-12, Fire Problems Program Topical Report 10 (Silver Spring, Md.: Johns Hopkins University, Applied Physics Laboratory, 1971).

9. Ibid.

10. B. M. Halpin, J. J. Dinan, and O. J. Doters, *Assessment of the Potential Impact of Fire Protection Systems on Actual Fire Incidents* (Laurel, Md.: Johns Hopkins University, Applied Physics Laboratory, 1977).

11. Unpublished paper by B. M. Halpin presented at Symposium, Johns Hopkins University, Applied Physics Laboratory, 1977.

12. B. M. Halpin, J. J. Dinan, and O. J. Doters

13. R. M. Fristrom and A. G. Schultz, *Annual Summary Report, July 1, 1970–June 30, 1971,* Fire Problems Program A71 (Silver Spring, Md.: Johns Hopkins University, Applied Physics Laboratory).

14. *Fire Death Study for 1981 in Alabama* (Montgomery: Alabama State Fire Marshal's Office, 1982).

15. Personal communication to the author from John S. Robison, Alabama State Fire Marshal, 1994.

16. *The Human Factor in High Fire Risk Urban Residential Areas: A Pilot Study in New Orleans, Louisiana* (Baton Rouge: Planning and Research Associates, Inc., prepared for the National Fire Prevention and Control Administration, 1976).

17. *Fire Prevention Program Planning in a Social Action Perspective: An Exploratory Study of Knowledge, Attitudes, and Leaders in a High Fire Risk Area of New Orleans, Louisiana* (Baton Rouge: Planning and Research Associates, Inc., prepared for the National Fire Prevention and Control Administration, 1977).

18. *The Human Factor in High Fire Risk Urban Residential Areas,* pp. 1, 2.

19. *Fire Prevention Program Planning in a Social Action Perspective,* p. 20.

20. U.S. Department of Commerce, National Institute of Standards and Technology, *NSIT Building and Fire Research Laboratory* (Gaithersburg, Md.: NIST, 1993),

21. National Institute of Standards and Technology, Building and Fire Research Laboratory, *Projects 1993* (Gaithersburg, Md., 1993), pp. 91, 98, 119.

22. National Institute of Standards and Technology, Building and Fire Research Laboratory, *Impacts* (Gaithersburg, Md., 1993), pp. 24, 35.

23. National Fire Laboratory of the National Research Council, Canada (Ottawa, 1993).

24. *Fire-Code Inspections and Fire Prevention: What Methods Lead to Success?* The Urban Institute and National Fire Protection Association supported by National Science Foundation and U.S. Fire Administration (Washington, D.C., 1978).

25. U.S. National Commission on Fire Prevention and Control Report, *America Burning* (Washington, D.C.: U.S. Government Printing Office, 1973).

26. R. A. W. Switzer and D. M. Baird, *Study on Fire Prevention and Control Systems in Canada* (Ottawa: National Research Council of Canada, 1980).

27. *Indian Conditions: A Survey,* Minister, Indian and Northern Affairs Canada (Ottawa, 1980).

28. Thomas R. Wood, *A Study of Structure Fires and Fire Sprinkler Systems in U.S. Fire Stations, An Applied Research Project Submitted to the National Fire Academy as Part of the Executive Fire Officer Program* (Emmitsburg, Md. 1992).

29. Ed Hayman, *Wood Shake Shingles . . . the 800 Year Old Problem, An Applied Research Project Submitted to the National Fire Academy as Part of the Executive Fire Officer Program* (Emmitsburg, Md., 1992).

CHAPTER 13

International Efforts in Fire Prevention

In any attempt to assess current fire prevention programs on this continent, it is desirable to compare our procedures with those in effect in other parts of the world. Some fire prevention practices followed in the United States and Canada can be linked to procedures in other countries of the world; however, there are probably as many differences as there are similarities.

Assumptions may be made that all countries follow the same fire prevention practices. This is not true. Procedures practiced in Italy, for example, are quite different from those practiced in Hong Kong. The fire preventionist can learn a great deal from a study of these various procedures.

In making comparisons among countries, many other factors must be taken into account in analyzing fire prevention practices and their results. Standards of living, industrialization, customs, dollar valuations, use of electricity, even smoking habits vary considerably from country to country. All of these factors, as well as a number of others, may influence the fire situation in a given country. An intangible element that makes a great difference is the general degree of adherence to constituted authority. A tightly controlled population where the day-to-day lives of citizens are closely watched by the government produces an entirely different fire potential than in a country where the citizens are free to act without close control. The availability of insurance is a factor; in some countries it is difficult to obtain. Travel restrictions within a country may also have a bearing on the fire situation. In some countries, travel from one section to another is severely restricted, which limits the possibility of an individual setting a fire in one community and escaping to another.

Heating methods employed are a major factor. In some countries, very few open-flame devices are available or used. In others, open-flame devices are quite common. There are great varieties in length of heating seasons. Practices in construction likewise vary considerably from place to place. In some countries, practically all construction is masonry or fire-resistive. In others, the common practice is to use wood, or even cardboard, for most buildings.

Fire prevention practices in a country where construction is primarily noncombustible are not the same as in one in which construction is primarily wood frame or plywood. Fire deaths, however, may not be as directly related to construction as is commonly thought.

At any rate, comparisons from country to country must be made with the previously mentioned factors in mind. Conclusions made without consideration of these factors may be invalid.

Fire Death Statistics for Various Countries _____

Before making a detailed study of fire safety procedures in other countries, it is desirable to review fire loss statistics from those nations. Again, in reviewing fire loss statistics, differences in valuations, standards of living, etc., must be considered. An interesting comparison may be made on a city-by-city basis as well. As an example of the country-by-country comparison on fire occurrences and losses, a comparison may be made between the United States and several other countries. It should be noted that procedures, problems, and fire loss statistics are generally similar in the United States and Canada.

Fire death statistics show the United States with the highest death rate: 18.5 fire deaths per one million population. This figure is based on estimates of the National Fire Protection Association. Roughly two thirds of those deaths occurred in a victim's own home.[1]

The United Kingdom's fire death rate is 15 per million. However, there is a drop to the next highest industrialized nation, Canada, with 13.9 fire deaths per million.[2] The Netherlands has 9.4 million; Austria has 7.6 per million.[3]

On a per capita basis, fire losses in Canada are very similar to those in the United States. Residential fires account for 85 percent of fatalities and 46 percent of dollar loss. The latter figure is slightly higher than that for the United States. Canada has had a comprehensive nationwide fire reporting system for most of the century, and comparative statistics are quite meaningful. Causes of fires are likewise fairly similar to those south of the border. As with the United States, there are considerable varieties among the 10 provinces and two territories. In 1991, the fire death rate ranged from .0 per 100,000 population in the province of Prince Edward Island to 15.6 in the Northwest Territories. The 10-year average for the Northwest Territories was 12.5, as compared with a low of 1.6 in Quebec. One major fire in a sparsely settled territory skews statistics for many years.[4]

Comparison of Number of Fires

In the United States, 2.5 fires per 1,000 people occur in buildings.[5] This compares with 0.9 per 1,000 in The Netherlands, and 0.9 in the United Kingdom.[6] Per capita fire losses also vary considerably from nation to nation; however, the United States and Canada remain near or at the top in most compilations.

These differences have existed for many years. Mr. C. A. Hexamer, President of the National Fire Protection Association in 1906, observed that in the previous year per capita fire loss in the United States had been $2.47, while the average per capita fire loss for six European countries for the same year had been 63 cents.[7]

Mr. Franklin H. Wentworth, Secretary, National Fire Protection Association, addressed the comparison with Europe in a series of talks in 1910. His comments included the following point:

> The European peoples point to the fire waste of America as evidence that the American people are the most careless and irresponsible individuals in the civilized world. This is not a flattering tribute. It behooves us to inquire if the charge be true, and if so, how we may mend our unpleasant reputation. Statistics show that in the United States and Canada the fire waste is roughly ten times as much per person as in Europe. This contrast is partly explained by the facts that there are more people in Europe upon whom to figure this percentage and that more buildings in America are constructed of wood. Outside of Constantinople with its unsanitary conditions, European cities are seldom visited by such sweeping fires as have devastated the American cities of Chicago, Boston, San Francisco, Atlanta, Baltimore, Chelsea, Salem, Nashville, Augusta, and others. American building construction is inferior to that of Europe in other respects. A poorly built city with numerous wooden buildings awaits only the right kind of fire on the right kind of night for its complete destruction.[8]

This condition also prevailed in 1930, as noted by Henry G. Hodges in *City Management*. Hodges stated, "The 1930 fire losses in the United States were greater than in all Europe combined. The number of fires per thousand of population in the average American city is six times as great as it is in European cities. With approximately the same population, Berlin's fire loss is one-tenth that of Chicago." Even then U.S. losses per fire, however, were $241 as compared with European losses of $668 per fire.[9]

Comparison of Losses per Fire

It should be noted that the monetary loss per fire is generally greater in other countries than in the United States and Canada. This is probably attributable to several factors. One of these is the fact that North American citizens are probably less reluctant to summon the

Population comparisons for fire deaths (1986-1988)

Country	Deaths per 100,000 persons	
Switzerland	0.44	
Netherlands	0.57	
Austria	0.81	
Germany, West	0.96	
Spain	1.15	(1986)
France	1.34	
Denmark	1.50	
Sweden	1.57	
Japan	1.60	(1985/1986)
Norway	1.64	
New Zealand	1.68	(1985/1986)
Belgium	1.78	
United Kingdom	1.81	
Canada	2.20	
Finland	2.46	
United States	2.63	
Hungary	3.25	

Figure 13-1 Chart showing comparative statistics for deaths due to fires. (World Fire Statistics Centre, London.)

fire department than is the case in some other parts of the world. There is no penalty imposed on an individual who calls the fire department in the United States and Canada unless it can be proven that it was a malicious false alarm.

In the United States and Canada fire stations are more closely spaced than in a number of other countries of the world, thus fire apparatus is more likely to arrive in a shorter period of time. In a number of countries, telephones are few and far between, which makes it difficult for a person to transmit a fire alarm. There are fewer fire extinguishing systems in most other countries than in the United States and Canada.

An interesting method of calculating fire losses is used by the Fire and Emergency Services Department of Johannesburg, Republic of South Africa. Monetary losses in fires are calculated in relationship to insurance risk. Each year the total number of building fires is listed, as is the insured fire risk: total building and its contents that carry insurance. The total actual fire damage in building fires is developed. A percentage of loss is then calculated, which usually runs well below 10 percent. An argument can then be made that the fire department was able to save 90 percent of the building. While there are loopholes in this procedure, there are times when the method is meaningful.[10]

Causes of Fires _____

A comparison of statistics on causes of fires can also be interesting. Variables in assigning fire causes, however, makes these figures less valuable than the previously mentioned comparisons because of classification. For example, reports of the London Fire Brigade[11] include as a cause of fire "structural defects" and "fires in course of business operation." Neither of these normally appears as a cause in the United States or Canada. They are contributing factors, possibly, but are not actually specifically thought of as causes of fires.

Primary causes of fires in Moscow are careless adults; careless children; heating installations and irons; followed by lightning and "others." In St. Petersburg, 45 percent of fires are caused by general carelessness; 18 percent by lightning; 13 percent by domestic electricity; and 12 percent by chimneys and stoves. Careless children cause 7 percent. An estimated 46 percent of St. Petersburg fires occur in domestic buildings, and 20 percent are in offices.[12]

A review of the Italian reports again shows the problem with attempting comparisons of statistics. Causes as listed by the Italian Fire Service, which incidentally is a nationalized fire service, include "heating, cooking, chimney, flues, ashes"; "electrical"; "unknown"; and "miscellaneous known causes." This grouping is quite different from that employed in many other countries of the world. The "miscellaneous known causes" classification could include causes that are specifically reported in other jurisidictions. It is difficult to make meaningful comparisons between these figures and those developed in the United States and Canada.[13]

Rural Fires _____

All figures quoted so far for the various countries include rural areas as well as urban areas. The variables that arise through inclusion of rural fires again color any comparisons that may be made. There is a wide difference among countries in providing fire service in rural areas. Some countries have practically no organized fire departments in rural areas; others have fire departments in practically every section. The existence of a department would affect the loss-per-fire figure, as well as the total fire loss figure. A 100 percent relationship cannot be assumed, however, because fires in areas without fire departments may in some cases not be reported. Without benefit of a fire reporting procedure, no one in an official capacity may be aware of the existence of burned-down buildings.

Variables that Affect Statistics _____

Variables from country to country, and even within a particular country, make it difficult to assess comparative fire loss figures. Following are some of the allowances that must be made in evaluating statistics.

Placement of Fire Stations and Equipment

In reviewing the total fire loss figures and the loss-per-fire figures, the placement of fire stations has been mentioned as a factor. Countries in Europe as well as in Asia generally have very few fire stations considering the area of the city, while cities in the United States and Canada maintain many more.

Equipment used in the control of fires is another consideration. European fire stations, especially those on the continent, are provided with a much greater number of units per station. It is not unusual for a continental fire station to have six or seven pieces of apparatus operating out of the same station. This arrangement is seldom found in U.S. and Canadian fire stations. The single-engine substation found in many North American cities is less common in Europe.

While the European system involves far fewer stations, a larger number of active fire fighting companies are found in each station. The arrival of these units at the fire scene at approximately the same time creates a different tactical situation than does the American system of having units arrive on the fire scene on a staggered basis, depending on the distance from the station. The American system usually provides for earlier application of water on the fire; however, difficulties may be encountered in getting major operations under way.

This has an affect on loss experience as well as on fire prevention practices. The neighborhood fire station concept should bring about a greater awareness of fire safety in a neighborhood. It stands to reason that more positive contact with the public can be made through the latter concept. Personnel assigned to a neighborhood station are aware of fire problems in the neighborhood. Members of a more remotely stationed company may not be able to develop this personal contact with citizens of the community they serve. Even in some heavily built neighborhoods, the fire station is looked on as an integral part of the community.

Construction Practices

As previously mentioned, construction practices vary from continent to continent and even within countries, states, and provinces. However, a higher percentage of fire-resistive and noncombustible construction is employed in Europe than in the United States and

Canada. European construction practices prevail in Central and South America. Even though a majority of European structures, including those for residential occupancy, are primarily of noncombustible outer wall construction, many have wood frame interior construction. This construction is intended to prevent conflagrations. The tenant, owner, or occupant will have a personal interest in fire safety, thereby mitigating results of fires from within. Experiences of the European countries during World War II further emphasized the need for fire-resistive outer wall construction. Conflagration potential is also considered.

The United States and Canada have a high percentage of combustible construction. This type of construction has been prevalent throughout the history of both countries and has been responsible for many major conflagrations. Included are many fatal fires in which construction was a factor.

Availability of materials is a factor in construction practices. In the United States and Canada, building methods also vary according to availability of materials within given areas. As an example, the ready availability of wood in the Pacific Northwest states has resulted in a higher percentage of wood frame construction than is found in eastern cities.

Occupancy Factors

A study of housing ownership may shed some light on fire experience. A fairly high percentage of people living in European countries die in the same house in which they were born. It is not uncommon for citizens of most countries of the continent to live their entire lives in the same structure.

The average American and Canadian, on the other hand, moves from dwelling to dwelling several times during a lifetime; studies indicate an average of seven moves, in fact. It is unusual to hear of people who die in the house in which they were born.

An individual who lives an entire lifetime in the same house is probably more apt to value the home and its contents than is a more transient individual. The stationary individual will probably take more adequate precautions against fire. Renters usually have less interest in home maintenance than owners.

The level of income necessary to purchase furnishings and other personal properties is less in this country and therefore decreases the significance placed on the loss of such materials in a fire.

Another factor is that of reliance on the fire department. With neighborhood fire stations, the American or Canadian may feel less resistant in summoning the fire department. Citizens consider that this is a public service to which they are entitled as taxpayers. A higher percentage of emergencies are reported to the fire department in the United States and Canada than in other countries.

On several occasions fire departments in the United States have had to make public announcements that they were unable to respond to alarms because of severe weather conditions. In recorded cases, fire alarms have dropped off to practically nothing.

In one case, in Milwaukee, emergency storm conditions lasted for over 48 hours, during which time it was practically impossible for apparatus to move out of the stations. The public reacted to radio announcements, and no serious fires occurred during the critical time. There is no way to know how long this restraint could be sustained. Indianapolis had a major flood in 1913 that lasted for four days. During this time normal water supplies were knocked out, and the fire chief announced that the fire department was helpless. No fires occurred during that time; however, they resumed when fire protection was restored.[14] Fire responses did not decrease, however, in Des Moines, Iowa, when the 1993 flood overtook their water works.

Authority in Inspections

The matter of authority connected with inspections undoubtedly has a bearing on the occurrence of fires in a given country. A general weakness in control through inspections and other methods of enforcement can result in a greater frequency of fires than when stringent requirements are vigorously enforced.

In an exchange visit, the fire chief from St. Petersburg, Florida, visited St. Petersburg, Russia, in 1992. He found that the fire department, which is operated by the national government, has a strong fire prevention program, which includes plan reviews, issuance of certificates of occupancy, and strict code enforcement, including the ability of a fire inspector to immediately close down a business found in violation. Fire inspectors are disciplined when a preventable fire occurs in their territory. The city of almost six million population loses about 200 people a year in fires.[15]

Comparisons of Fire Prevention Education and Enforcement _____

Emphasis in the field of fire prevention varies from country to country. It is interesting to note that the United States and Canada apparently rely more heavily on fire prevention education than do some other countries.

In Japan, Tokyo's fire department is an example of a department with well-organized education and enforcement activities. All plans for construction must be approved by the fire department prior to the issuance of a building permit. Fire fighters check structures to assure compliance during and at completion of construction.[16]

Spring and autumn fire prevention campaigns are observed each year. These include fire drills, demonstrations, and fire prevention meetings. A mobile disaster protection classroom travels over the city to educate the public in disaster protection, including fire prevention.

Fire protection equipment in buildings is inspected on a regular basis by fire service personnel. Larger buildings are required to have fire protection managers who are trained by the fire department.

Closely related to fire prevention are the public relations activities of the department. These include public fire education, and arranging for fire service personnel to visit elderly people who are "shut-ins" to assist them with fire precautions. Of the 18,000 members of the Tokyo fire department, 1,850 devote their full time to fire prevention duties.

Philip Schaenman examined fire practices in Japan. He noted that many people live in wood and paper houses with very narrow gaps between houses. He states that:

> The Japanese focus on educating the public in fire-safe behavior, extinguishment of small fires by citizens using buckets of water or extinguishers, and escape practices. There is tremendous social pressure to be fire safe. Virtually the entire population is reached by local fire safety education programs starting in the home and preschool, and continuing through elementary school, junior high school, on the job, and on the street. Neighbors practice fire safety drills together. The Japanese experience demonstrates that public safety education alone can reduce fire incidence enormously.[17]

European countries strongly enforce fire safety regulations. In West Germany and other European countries, chief officers of the larger fire departments are for the most part graduate engineers. This ensures a high level of technical guidance in matters of fire suppression and gives the department a technically competent voice in the development and enforcement of fire safety standards. These chief officers do not as a rule come through the ranks of the fire brigade but enter directly into the administrative level of the fire brigade. These individuals may transfer from one jurisdiction to another, much the same as with city managers in the United States. This two-tier system is much like that of the military services in the United States and Canada.

Fire safety activities of the London Fire Brigade are described as follows:

> The London Fire Brigade attaches great importance to their community programs striving to "prevent" rather than "cure" fire. Over 100,000 visits are made each year to public buildings such as theaters, offices, shops, factories, hotels, schools, health care establishments and petroleum installations. Over 15,000 sets of plans are examined as well. The London Fire Brigade also speaks to over 70,000 children between the ages of 9 and 11 each year.

Two areas of particular emphasis in the Brigade's safety campaigns are the value of installing smoke alarms and the inherent dangers of foam-filled furniture. At the urging of the country's fire brigades, Parliament implemented a restriction on the use of dangerous foam in furniture in 1989.[18]

Fire services in the United Kingdom are organized on a countywide basis except in Scotland and Northern Ireland, where a regional concept is used. Both full-time and part-time personnel are employed. This system provides full-time fire prevention personnel throughout the country, even in less populated areas protected by part-time or volunteer fire fighters.

Another noteworthy feature of the British fire protection system is the required use of fire safety rating signs in hotels and other places of public accommodation. A guest registering at a hotel is thus aware of safety factors in the structure. The property must be closed if the rating falls too low.

The National Fire Service College for the United Kingdom offers many courses relating to fire prevention. A unique program is a one-month class for architects, engineers, and building officials in which the students engage in live fire training in addition to the study of fire protection systems and related subjects. This college is considered to be the finest such facility in the world.

Construction practices favor limited fire spread in Central and South America. Climatic conditions, especially in Central America, preclude heating-system fire problems. These countries, however, are finding that fire protection and fire prevention problems are rising with the development of new technology. As an example, the popularity of liquefied petroleum gas as a fuel for heating and cooking has introduced a new fire protection problem in some of these countries.

Variations in Fire Code Applications

A major difference in philosophy exists between the United States and other countries. The United States is almost unique in having a system in which codes vary from one jurisdiction to another. In most other countries, fire safety standards, whether strong, moderate, or weak, are uniform within a given country. There may be minor differences from city to city; however, the general application of the code is usually the same throughout a given country. Application is much more uniform in Canada than in the United States.

In the United States a city may have a very strong fire prevention building code, yet the unincorporated areas of the same county may have no code. There are many examples of hazardous operations that would not be permitted within the city limits but are carried out with no problem in adjacent unincorporated areas.

Building construction, sales of hazardous materials, and dispensing operations vary considerably across municipal boundaries throughout the United States. A visitor from another country has a difficult time understanding why a hazard is treated differently on one side of a municipal boundary than on the other because the immediate surroundings, people, activities, etc., appear to be the same.

Variances in Practices and Procedures

In the United States there are examples of variances in practices and procedures among states. These differences appear not only in fire prevention and building codes, but in other fire-related measures. As an example, one municipality may have a program for fire safety training for schoolchildren that is not conducted in neighboring communities. It is a matter of chance as to whether or not the school-age child has the benefit of fire prevention education in the classroom. The National Commission on Fire Prevention and Control looked on this variety in fire prevention efforts as a major problem.[19]

National enforcement agencies are usually responsible for similarities in code enforcement procedures in other countries of the world. In many nations, the national government does not directly enforce fire prevention requirements but instead supervises local enforcement procedures. The national government enters the case only when there is evidence that the local agencies are not performing satisfactorily.

Fire prevention practices and enforcement procedures generally reflect the history of the nation as far as governmental control is concerned. Fire prevention practices in the United States probably are not uniform partially as a result of the experiences and attitudes of those who founded our government. Many immigrants came to North America in an effort to get away from extreme governmental regulation. As a result, the only requirements imposed by the original government in this country were those considered absolutely necessary for survival.

As urban areas developed, fire prevention and building code requirements became necessary to ensure domestic tranquility and safety. Individuals living in less densely populated communities saw no necessity for regulations to control conditions they did not view as hazardous. Therefore, we find that a great deal of the fire prevention and building code coverage in the United States came about as a result of specific problems arising at given times. Many states still lack fire prevention and building codes covering all conditions that might exist within the state. An example is the control of liquefied petroleum gas by a state that does not deem it necessary to maintain regulations governing overcrowding of places of assembly.

Fire Costs

A summary of fire costs from a number of countries, primarily in Europe, is included in the following report from the World Fire Statistics Centre:

> The World Fire Statistics Centre, Geneva, Switzerland, publish annual estimates of the national fire costs of over 15 countries, mainly in Europe but including USA, Canada and Japan. The figures are broken down into seven items of fire costs, first analyzed by Tom Wilmot in his report, "European Fire Costs—The wasteful statistical gap," published in 1979 by the Geneva Association. These costs are: direct losses 30 percent; building protection 30 percent; fire brigades 15 percent; fire insurance 15 percent; indirect losses 5 percent; human losses 5 percent.
>
> Wilmot found that the total fire costs run at a rate of roughly 1 percent of the Gross National Product. The main conclusion of his report was that little political will existed in Europe to cut these costs. This lack of will was reflected in the paltry national efforts made to produce reliable fire statistics, which are needed to draw public attention to the size of the problem and the need to reduce the costs. He concludes that compared with road safety, crime prevention and industrial safety, fire prevention signally fails to receive the high level political attention that it merits.
>
> Since 1979, the Centre has worked towards attracting greater political interest in the need to cut national fire costs. The message has been delivered at international fire seminars, including regular Centre Seminars in Geneva and European Commission Seminars in Luxembourg. Close contact is maintained with the European Union in Brussels at political and civil servant levels. The problem of Arson statistics is being tackled jointly with the European Arson Prevention Institute. Reports on the national fire statistics are made regularly to the United Nations.[20]

The extensive studies reflected in this report represent a thorough analysis of the complete fire situation in Europe and several other countries. It is easy to see that effective fire code enforcement, building construction requirements, and public fire safety education can have a positive effect on reducing the percentage of the Gross National Product that is consumed by fire-related costs. These resources can then be used more productively.

Industrial fire losses constitute a high percentage of monetary losses in the Republic of South Africa. Paper and printing, chemicals and petrochemicals, and electronics are among the major contributors to this condition. The Fire Protection Association of Southern Africa, a counterpart to the National Fire Protection Association, is waging a major effort to reduce fire losses in that nation. The Association operates a fire college, which specializes in training industrial employees.[21]

Philip Schaenman's report, *International Concepts in Fire Protection: Ideas from Europe that Could Improve U.S. Fire Safety,* suggests that some European fire safety procedures are worthy of emulation in the United States and Canada.[22] The author visited a number of European countries in an effort to learn why fire statistics are so much more favorable on the other side of the Atlantic Ocean.

The report suggests that European citizens are more aware of the hazards of fire than are North Americans because of the major fires that have stricken their cities during wars and during the Middle Ages. They support stronger fire regulations and receive more instruction on fire safety from their parents. The report also suggests that these countries have stronger building and fire safety codes and that these codes are in effect in rural as well as urban areas, while many rural areas in the United States lack such codes. (Insurance also plays a more prominent role than in the United States. Insurance carriers have plan review authority and may, in fact, operate as government agencies.)

The extensive use of chimney sweeps in Europe may also help prevent fires in residences. In some countries, visits by chimney sweeps are mandatory. Consumer products are regulated more strictly, and electrical and gas-powered products cannot be sold unless tested and approved by the government.

Schaenman found that European fire officers generally have more education in fire protection and building technology than their counterparts in the United States. This means that they are more oriented toward plan review and code enforcement. He also found that fire fighter injuries are few and far between in Europe. Self-contained breathing apparatus has been widely used in Europe since well before World War II. Fire fighters have ridden inside vehicles rather than on the tailboard for an equally long period of time.

Most European fire fighters are trained in fire prevention and code enforcement in their recruit training programs, while such training is less common in the United States. Senior fire officers in Europe devote a much greater percentage of their time to fire prevention and code enforcement duties.

The report also notes that the Fire Research Station at Borehamwood, United Kingdom, has been switching its research emphasis from structures toward contents, reflecting a growing need to address fire problems related to consumer products rather than to construction features.

Schaenman has prepared a third report, *International Concepts in Fire Protection—New Ideas from Europe.* Fire safety efforts of the United Kingdom, The Netherlands, Austria, and Hungary are examined in this publication. Austria and The Netherlands are recognized as having among the lowest fire death rates in the world. Both have excellent fire safety education programs.[23]

Summary

In making comparisons of fire prevention programs in effect on this continent with those in other parts of the world, there are probably as many differences as there are similarities. Standards of living, customs, dollar valuations, use of electricity, even smoking habits vary considerably from country to country. An intangible element that makes a great difference is the general adherence to constituted authority in many other countries. Travel restrictions, different kinds of heating devices, and methods of building construction are further examples of factors that must be taken into account in analyzing fire prevention practices and comparing results.

Some of the variables that affect statistics are diversified emergency services in the various countries; differences between the North American and European fire service systems with respect to placement and number of fire stations and amount of fire control equipment at each; and the greater use of noncombustible outer wall construction in European countries.

Psychological factors also have a bearing on the fire situation in a country, and more research in this area is needed. It is not uncommon, for example, for Europeans to live out their entire lives in the same houses in which they were born. These individuals would be more apt to value the home and its contents and take greater precautions against the threat of fire than someone who moves about from dwelling to dwelling. The weight of authority in inspections also undoubtedly makes a difference in the occurrence of fires in a given country.

A major difference in philosophy exists between the United States and other countries with regard to the application of fire codes. In most other countries, fire prevention and building codes are as a rule uniform within the country and are usually enforced by a national agency. In the United States, many variations in code applications exist, in neighboring municipalities as well as from state to state, and enforcement is often left to the jurisdiction concerned. Canada's procedures are much more uniform than those of the United States.

Comparisons of our fire prevention practices with those of other countries will provide us with some valuable information. We can, in fact, learn a great deal from such studies and may find that it is highly desirable to include some of the better features of foreign programs in our own.

Notes

1. Personal communication to the author from the National Fire Protection Association, 1994.
2. Personal communication to the author from the Fire Commissioner of Canada, 1994.
3. Philip S. Schaenman, *International Concepts in Fire Protection—New Ideas from Europe* (Arlington, Va.: TriData Corp., 1993), pp. 9, 33, 51.
4. Personal communication to the author from the Fire Commissioner of Canada, 1994.
5. Personal communication to the author from the National Fire Protection Association, 1994.
6. Philip S. Schaenman, *International Concepts in Fire Protection—New Ideas from Europe.*
7. Percy Bugbee, *Men against Fire: The Story of the National Fire Protection Association, 1896–1971* (Boston: National Fire Protection Association, 1971), p. 15.
8. Ibid., p. 19.
9. Henry G. Hodges, *City Management Theory and Practice of Municipal Administration* (New York: F. S. Crofts and Co., 1939), pp. 486–488.
10. Annual Report of the Chief Officer Fire and Emergency Services Department (Johannesburg, Republic of South Africa, 1983).
11. Boris Laiming, "Twentieth Century Fire Protection: History, Highlights, and Trends," mimeographed paper, Table 4.
12. D. I. Lawson, "Fire Protection in the USSR," FPA Journal no. 64, Supplement 1 (London: Fire Protection Association) pp. 8, 9.
13. Boris Laiming, "Twentieth Century Fire Protection, History, Highlights, and Trends," Table 5.
14. Percy Bugbee, *Men against Fire,* pp. 35, 36.
15. Jerry G. Knight, "From St. Petersburg to St. Petersburg," *Fire Chief* (August 1993).
16. *White Book on Fire Service in Japan* (Tokyo: Japan Fire Protection Association, 1982).
17. Philip Schaenman and Edward Seits, *International Concepts in Fire Protection: Practices from Japan, Hong Kong, Australia and New Zealand* (Arlington, Va.: TriData Corp., 1985).
18. Jerry Senk, "The London Fire Brigade," *The Voice,* (July–August, 1992).
19. U.S. National Commission on Fire Prevention and Control, *America Burning* (Washington, D.C.: U.S. Government Printing Office, 1973).
20. Tom Wilmot, *World Fire Costs* (London: World Fire Statistics Centre, 1994).
21. Peter Davey, "Fire Statistics," *Fire Protection,* Marshalltown, South Africa, Fire Protection Association of Southern Africa (June, 1991).
22. Philip S. Schaenman, *International Concepts in Fire Protection, Ideas from Europe That Could Improve U.S. Fire Safety* (Arlington, Va.,: TriData, 1982).
23. Philip S. Schaenman, *International Concepts in Fire Protection New Ideas From Europe.*

CHAPTER 14

Proving Fire Prevention Works

Fire prevention works. However, it must be proven year after year. The preceding chapters have illustrated many examples of successful fire prevention programs.

In 1928, Chief W.D. Brosnan of Albany, Georgia, spoke on this subject at the first annual meeting of the Southeastern Association of Fire Chiefs. With the change of a few words, the talk would be as appropriate today as it was in 1928.

> Any person who is at all conversant with fire safety knows that at least eighty-five percent of [fires] could be prevented. It is the duty of the Fire Chief to assume leadership and point out the way for the protection of life and the conservation of property of our citizens.
>
> If the fire loss of the country is to be reduced, we must get away from the out-of-date methods of the old time red-shirt brigade. They thought the duties of the firemen were to sit around the engine houses waiting for an alarm of fire and then proceed to extinguish it as best they could; but the modern Fire Chief knows that he must be up and doing and prevent fires from starting, if he is to be successful in reducing the loss.

Chief Brosnan highlighted the reduction of fires in the previous five years. The following reduction occurred despite a 40 percent population increase: 1923, 160 fires; 1924, 159 fires; 1925, 125 fires; 1926, 110 fires; 1927, 97 fires.

Albany maintained a night patrol by uniformed fire fighters. These individuals were detailed from fire companies and were required to look into every mercantile occupancy once each hour. In 1927, this patrol discovered more than 85 percent of mercantile fires, thereby keeping losses at a minimum.[1]

It must be recognized that a tremendous amount of progress has been made in fire safety and prevention. The popularity of smoke detectors and residential sprinklers is an example. Public fire safety education is an accepted program of many fire departments. Thousands of citizens of all ages are still not fortunate enough to have the benefit of this important information.

Total Tax Dollar Loss
1981 - 1991
For Fires Occurring Between 1981 and 1985

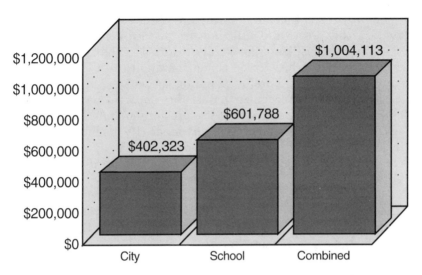

For Fires Occurring Between 1986 and 1991

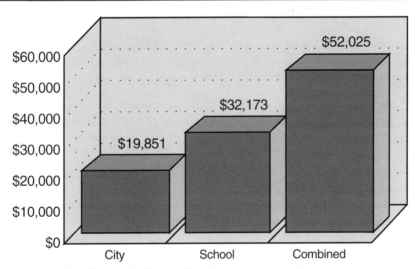

Figure 14-1 The Mansfield, Texas, Fire Department used these graphs to illustrate the city's reduction in tax dollar loss subsequent to the establishment of a fire prevention bureau in 1986. (Adapted from the report prepared by Mansfield Fire Department Administrative Staff, 1992.)

Great strides have been made in hotel–motel fire safety. Federal legislation mandating that government employees on official travel stay in fire-safe properties is a major step forward in this field. There are many other examples of fire safety improvements in publicly occupied buildings.

Offsetting some of the progress in fire safety is the reduction of the staffing of fire prevention bureaus in many cities as an economy move. The widespread practice of utilizing fire service personal for emergency medical duties has undoubtedly resulted in the saving of hundreds of lives, but it has forced some communities to eliminate fire company inspections and prefire planning surveys.

Fire service leaders recognize that the fire department alone cannot successfully impart the message of fire prevention and safety without strong community involvement. In many inner-city neighborhoods and low-income rural areas, individuals who are known and respected by the residents may be quite effective in generating an interest in fire prevention.

Fire prevention chiefs must be able to adjust their programs to meet new challenges such as the urban–wildland interface problem and the continuing expansion of the use of hazardous materials. Bureaus incapable of change are doomed to failure.

Fees for Services

Some cities collect permit fees for certain fire prevention services. The rationale for the advent of this concept is explained in the following memo from the Western Fire Chiefs Association, Inc.:

> Fire departments have historically been emergency services organizations and have given their services to the public free of charge. In recent years governmental leaders have questioned the high cost of providing fire protection and have looked for ways to off-set these costs.
>
> On the other hand, buildings departments for years charged various fees for building permits, plan review, and inspections. In many communities the building departments actually turn a profit and generate more revenues than it costs to operate their departments. The public and private sectors accept these fees as part of the cost of doing business.
>
> Fire departments have been involved with plan reviews and building inspections where the building department was collecting fees and the fire department performed similar services at no charge. This has led to the conclusion by many fire departments that they should share in plan review and inspection fees. Also that individuals or firms that engage in activities that require permits under the code, should be required to pay for the special services required for the issuance of the permit.

In some communities the fee permit concept has been successful because the business community was made aware of the program and understood the intent from the time of implementation. In other cities the program had a very negative effect because the business people were uninformed and viewed it as just another unnecessary method of taxation.[2]

Measurement of Fire Prevention Effectiveness _____

Measuring the success of fire prevention efforts is subject to many variables.

As an example, a nationally circulated magazine attempted a comparative evaluation of fire departments through the use of per capita fire losses. Cities over the country were rated as to their fire departments' efficiency according to this criteria. The fire department protecting a bedroom community had a definite advantage over a community in which there was a good deal of heavy industry with little residential area.

Fire deaths totals have also been used as a means of measuring the effectiveness of fire prevention activities. Although such a measurement may have some validity, there are weaknesses in it as well.

Research has shown that socioeconomic factors have a bearing on fire safety measures and that a fair comparison of individual procedures in a community should take these into account. As an example, most cities experience a higher per capita fire death rate in low-income, row-house areas than in high-income, detached house areas.

Another measurement of fire prevention effectiveness is the relative position of the municipality from a fire insurance rating standpoint. The insurance rating classification does not directly reflect loss, except in major cities and counties.

Reports of achievement are also prepared at the national level. The U.S. Fire Administration convened a group of fire safety leaders in 1987 to review progress made in the field since issuance of *America Burning* in 1973. The report of this workshop, entitled *America Burning Revisited,* includes the following summary of achievement:

The annual number of fire deaths has decreased by 23% from 1975 to 1985 (an average annual reduction of 1900 deaths). On a cumulative basis, this reduction means that an estimated 6,900 lives have been saved between 1975 and 1985. Fire fighter deaths also have been reduced significantly.

Even greater reductions in fire deaths have been achieved within special categories. Clothing fire deaths have fallen by 73% over the period 1968–1983. In addition, children's clothing fire deaths have dropped by 90%.

The number of fires reported to fire departments has decreased by approximately 20% over the period 1975–1985. The number of reported fires has been decreasing even though the population is increasing which means that the number of fires is declining even on a per capita basis.[3]

Annual Report

Many fire departments prepare an annual report to give community residents and the official governing body an accurate picture of the fire department's operation. The annual report should be simplified so it can be readily understood by members of the general public. The report should be attractive and easy to read. The use of photographs and charts will improve the appearance of the publication.

The annual report presents an opportunity to get across specific messages over and above mere statistical reporting of activities. The annual report of the fire department of Atlanta, Georgia, for example, lists the number of inspections, number of fires, and dollar losses for each type of occupancy. The reader can compare inspection frequency with loss in each type of building (office building, restaurant, lumber yard, etc.). The report also compares fires and losses for the previous 10 years. Some reports include a detailed list of violations discovered during inspections. This is a good checklist for designing fire prevention activities.

A number of communities do not prepare annual reports because of the cost involved. A tight budget should not preclude the issuance of a report. Even a photocopied report can be successful in advising the public of fire department activities. Arrangements should be made to make the annual report available to the public. It may be desirable for the report to be made in two forms—one a more complete report for use by public officials, the other a summary report for distribution to citizens.

Often included in an annual report is a mission statement and review of activities. An excellent example of a mission statement and review of activities is that prepared by the Fire Prevention Bureau of Mount Prospect, Illinois, Fire Department. Six persons are assigned to the Bureau; however, shift fire fighters assist by inspecting low-hazard occupancies and in presenting fire education classes to children and senior citizens.

MISSION STATEMENT

The Mount Prospect Fire Prevention Bureau is established to provide quality fire prevention inspections and programs to the residents and workers in the Village of Mt. Prospect in the following areas: code compliance and enforcement inspections to all licensed and unlicensed buildings in Mt. Prospect; fire prevention/safety education for preschool children, school children, adults, and senior citizens in Mt. Prospect; plan review for new construction; false alarm management; fire investigation for cause and origin; and provide fire guards for public events. These services are performed by highly trained civilian and sworn personnel in a cost effective manner to make Mt. Prospect a fire-safe village in which to live and work.

ACCOMPLISHMENTS FOR FISCAL 1992/1993

1. Completed reviews of Village Codes and upgrade code review to 1990 BOCA Building and Fire Prevention Code.
2. Completed inspections of 2,300 occupancies and facilities in Mt. Prospect.
3. Completed fire protection plan review of 249 facilities building and/or remodeling in Mt. Prospect and on-site inspections to determine construction compliance.
4. Investigated 25 fires, inspectors called to determine the cause and origin of these fires.
5. Presented fire safety education programs in the schools, churches, and high-rise multi-family units of Mt. Prospect for a total contact of 24,493 individuals.
6. Inspected the removal and installation of 39 underground storage tanks.
7. Tracked annual sprinkler system certification to have all sprinklers certified annually (242 systems).
8. Monitored all false alarms and sent notices to multiple alarm offenders to repair systems.
9. Provided fire guards to all public gatherings required by Village Ordinance in Mt. Prospect; Lions Carnival; Park District Fair; Prospect High School events; and Jaycee's Carnival.[4]

Recording Fire Safety Activities _____

The fire reporting system should also include information regarding activities of the fire prevention bureau, including demonstrations, lectures, other public education programs, plan reviews, and all other activities that contribute to the goal of reducing fire losses. Any public contact has a bearing on the official function of the organization and should be included.

The training division of the fire department should maintain complete records on time devoted to fire prevention training. The training division may call on the fire prevention bureau for assistance in such instruction.

A fire department record system should include information relating to all responses made by the department. These responses should be classified and enumerated in such a manner as to give an accurate picture of services performed by the fire department.

Fire Company Records

In fire suppression, record keeping begins at the company level. The company officer, whether in an engine, ladder, or rescue company, is the individual primarily responsible for preparing reports. Records maintained at this level are often filled in under adverse conditions. Simplicity and uniformity must be major considerations. A separate report should be maintained on all inspections, prefire planning visits, and other activities carried out by the fire company.

Fire Prevention Inspection Records

Fire prevention bureau records should include a record of all inspections including each building subject to inspection by the department. Because the ownership and occupancy of a structure often change, problems may be encountered in attempting to maintain records of inspections under the names of owners or occupants. The owner or occupant may be listed on a cross-file system with the primary inspection record maintained by street address. The inspection record of a building should include the findings of each inspection and the disposition of requirements imposed as a result of these inspections. The file should also include a diagram of the structure, which might be used in the department's prefire planning program. Computers are very helpful in this process.

The file should include photographs of the building and of conditions found within it. Photographs and diagrams can be used to pinpoint the location of specific problem areas.

Copies of any orders issued in connection with the occupancy should be included in the file. Newpaper clippings regarding the occupancy can also be helpful. A news item may indicate policy changes by the operators of the structure and give other information that can be of great value in the event of a fire, such as changes in business conditions that might relate to the cause of a fire. Fire prevention bureau records are considered public records in most states. Subrogation, arson cases, as well as other litigation in which details of fires must be recounted make the maintenance of adequate and complete records mandatory. It is necessary to be able to immediately recount inspections conducted and to have information available regarding fire prevention education programs carried out by personnel.

Recording Fire Deaths and Injuries _____

Fire department records should include facts about injuries sustained in fires as well as about deaths incurred as a result of fires. Many communities do not have complete records, because in most cases they record only deaths and injuries in which fire department response was involved. Medical reports are usually not public records.

A death or injury by fire in which the fire department was not called and the victim was taken directly to the hospital or mortician may be unknown to the fire department. This is often the case for burn injuries in which the victim is transported for medical treatment by a relative or neighbor and no fire department emergency medical or fire response is involved. Fire departments need to maintain close ties with hospital emergency rooms and the medical examiner or coroner's office.

Statistical reporting of deaths and injuries related to fire has another inherent obstacle. There are many borderline cases in which it is difficult to determine whether the death or injury should be attributed to fire. The question of assignment arises in traffic accidents in which the vehicle burns. A question as to whether the death or injury should be counted as a fire death or as a traffic death often arises.

Recording Loss Statistics _____

Widespread involvement of the fire service in the provision of emergency medical service has necessitated the development of data systems to record these responses. Many of these systems function at the state, as well as local level. Motor vehicle injuries and fatalities, falls, cardiac arrests, drownings and other non-fire related incidents are recorded in connection with fire service response.

The availability of statistics on non-fire related responses has been the driving force behind the expanding involvement of fire services in educational programs aimed at abating these injuries and fatalities. Fire departments may be joining with other community departments and organizations in disseminating information designed to lower the frequency of such incidents. Examples include programs promoting seat belt usage, CPR training, pool safety, and fall prevention.

The fire department record system should include information on fire losses in the community. Included should be data relating to causes of fires, buildings and occupancies in which fires occur, and information related to losses incurred in these fires.

Obtaining loss figures for all fires is a problem because there are so many variables in developing such information. Many departments use loss statistics developed by fire department personnel. Others use figures developed by insurance loss adjusters. In some cases the departments only count losses not covered by insurance carriers. They believe that any dollar loss covered by insurance is not actually a fire loss and, therefore, does not count in their statistical report. Some fire departments classify losses as moderate, medium, high, or in other terms of a general nature. This "broad range" procedure makes it easy for personnel to complete their reports on the fireground; however, accuracy is not served by this practice. Personnel assigned to estimating losses need training in evaluation methods.

Records and reports are quite helpful in determining the value of fire protection devices such as automatic sprinklers and smoke detectors. Statistics have authenticated the value of automatic sprinklers; however, smoke detectors are rapidly coming of age in this respect. Many communities are recording all smoke detector operations in an effort to prove the value of the device. Statistics on performance coupled with examples of successful incidents in which smoke detectors saved lives can be of great assistance in promotional campaigns by fire departments.

National Fire Incident Reporting System _____

The National Fire Incident Reporting System is operated by the Fire Data Analysis Branch, a part of the U.S. Fire Administration. This system is designed to collect and analyze fire incident and casualty data on a nationwide basis. The standard *Uniform Coding for Fire Protection,* National Fire Protection Association No. 901, is used as the basis for the system.[5] Forty-four states are now in the system, as are 29 metropolitan areas. The U.S. Fire Administration is using the data for producing national estimates and for developing a data base for answering information requests from government, industry, and the public. The data center is feeding national summaries back to the states for their use in evaluating and improving fire protection programs.

Data for the system, including fire incidents and casualty reports, are collected at the state level from local fire departments. Statewide summaries are prepared, as are feedbacks to local fire departments. Statewide data tapes are sent to the national system.

The Oregon State Fire Marshal's Office has operated an excellent fire reporting system for many years. It includes a great deal of information not normally included in fire reporting systems. Detailed analysis of insurance loss data is included, as are statistical data on fire department activities in the fire prevention field. Inspections by

Deaths from Fire and Flames by Age Group
Mississippi Residents, 1992

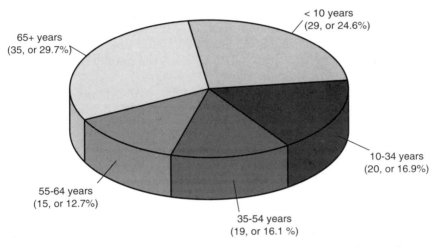

65+ years
(35, or 29.7%)

< 10 years
(29, or 24.6%)

10-34 years
(20, or 16.9%)

55-64 years
(15, or 12.7%)

35-54 years
(19, or 16.1 %)

Figure 14-2 Chart used by Mississippi State Fire Marshal to emphasize the high fire fatality rate among children and elderly citizens. (Public Health Statistics, Mississippi State Department of Health.)

occupancy, fire prevention programs, and permits issued and approved within each community are totalled in the report as well.[6]

It is imperative that fire incident data, whether gathered on a local, state, provincial, or national level, be of value to the agency submitting the initial report. Fire department personnel, whether career or volunteer, rapidly lose interest in supplying accurate information to a system that provides their department with few or no returns.

Use of Computers

Microcomputers are now widely used in fire prevention bureaus. The information age has provided the computer as a means of expediting decisions in fire safety code enforcement. The fire prevention chief is able to immediately review inspection criteria as well as results of code enforcement activities.

The term *computer* is derived from the Latin verb *computare*, which translates as "reckon" or "think." A computer is a mechanical device capable of thinking or reckoning. The term *microcomputer* refers to the smallest computer with full-function capabilities. Even one-person fire prevention bureaus are now taking advantage of this technology.

Computers reduce all information to simple electrical signals. This is accomplished regardless of the complexity of the information. A computer has a memory component, a control component, and an arithmetic component, as well as input and output capabilities. The memory component may be referred to as the storehouse for the computer. The actual physical portions of a computer are referred to as the hardware. Instructional systems that enable the computer to carry out its functions adequately are known as software.

Several general applications are available with software. The most popular is that of word processing, a process through which all types of documents—form letters, correspondence, and invoices—may be developed, revised, edited, and printed. Fire safety survey reports are being prepared with word processors in fire departments throughout the United States and Canada.

Another type of software provides for electronic spreadsheets. These may be used in much the same manner as the records kept by an accountant or bookkeeper. Budget totals and subtotals can be readily computed, as can hydraulic calculations for automatic fire extinguishing systems and water supplies for fire protection.

Data base management programs are available with the appropriate software. These may be used to store and retrieve sizable amounts of similar data, such as records of inspections, hydrant flow test data, and fire extinguisher maintenance records.

Computers may be arranged to transmit data from one to another using satellites or telephone lines. Fire service information is available through several national systems using this phenomenon. In some cities, microcomputers in fire stations link the entire department through a telecommunications computer network. Routine correspondence, newsletters, messages regarding hydrant and street closings, duty reassignments, inspection assignments, and other routine messages are handled in this manner.

Graphics can be developed with the appropriate software. Graphics may be used in annual reports, budget presentations, and other projects where visualization is important.

Computers play a major role in fire service management. These devices have enabled fire service managers to immediately ascertain trends and directions that may be significant when planning for fire safety within a community. Likewise, they play a major role in fire protection research. A computer can give the researcher accurate information on probabilities that may be difficult to emulate in the laboratory.

In many fire departments all fire fighters are proficient in the use of microcomputers. Operational requirements demand that this degree of proficiency be reached, if inspection programs, company response reports, company budget maintenance, and other essential activities are routinely processed by computer.

Computer-aided dispatch is becoming routine. This system usually includes methods of tracking the status of all fire fighting resources, occupancy data for hazardous and target locations within the city, and information on fireground operational requirements. Terminals for receipt of necessary data may be provided on fire apparatus and in command vehicles.

The best example of statewide use of computers for fire code enforcement is in the Commonwealth of Kentucky. The state has developed a program through which all Kentucky fire departments may report inspections and fires through 10 regional centers, which in turn are relayed to the state headquarters. The state also records and prepares reports to property owners for inspections made by local fire departments. Each department is given a printout with information on the status of all local fire inspections. The jurisdiction receives copies of the reports and is able to see the number and types of deficiencies within the community on receipt of each printout. This enables the fire prevention officer to keep track of inspections and to make appropriate personnel assignments from month to month. Many cities in the United States and Canada use a similar system for their fire inspection reporting and inspection frequency control. Other routine activities of the fire prevention bureau can be handled in a similar manner through the use of computers. Some track whether the person involved in a fire had received fire safety training.

Summary

Records are essential for the proper administration of any fire prevention program. Accurate and complete information regarding fire prevention and fire control activities should be available if a thorough job is to be done. Analysis of statistical records can reveal specific problem areas toward which fire prevention education programs should be directed. Analysis of records can likewise point up the need for changes in fire prevention codes, modification of statutory requirements, and other changes in regulations.

Examples taken from records of the local fire department are far more meaningful than are reports of fire conditions taken from communities hundreds or thousands of miles away. Last, statutory requirements make it mandatory to maintain records of information necessary in subrogation cases, arson cases, and other legal proceedings; and court cases in which fire departments have been sued make it imperative to keep records of all fire inspection activities and fire prevention education programs.

Records and reports of fire suppression activities and fire prevention efforts are closely allied, and it is sometimes difficult to draw a line between the two. The types of reports and records the fire department should generally maintain are a complete record of fire department responses; statistical records of fire deaths and injuries and fire losses; records of fire prevention and training activities; reports for public information.

A complete record of all fire prevention inspections is all-important in the work of the fire prevention bureau to ensure proper follow-up of requirements imposed or changes necessary to conform to fire code regulations. This file should be kept up-to-date with any new information acquired.

From all these records, many fire departments compile an annual report that gives community residents and the official governing body an accurate picture of the fire department's operation and accomplishments.

Notes

1. Southeastern Association of Fire Chiefs, *Proceedings of First Annual Convention,* (Atlanta, Ga., 1928), p. 18.
2. Memorandum to Montana Vice President, Western Fire Chiefs Assoc., from Uniform Fire Code Coordinator, 1986.
3. U.S. Fire Administration, Federal Emergency Management Agency, *America Burning Revisited* (Washington, D.C.: U.S. Government Printing Office, 1990), p. 141.
4. Village of Mount Prospect, Illinois Fire Department Fire Prevention Bureau, *Fire Prevention Report 1986 through June 1, 1993— Accomplishments* (Village of Mount Prospect, Ill., 1993), p. 57.
5. National Fire Protection Association, *Uniform Coding for Fire Protection,* NFPA 901 (Quincy, Mass., 1990).
6. Office of State Fire Marshal, *Annual Report for Calendar Year 1989* (Salem, Ore., 1990).

Index